Student Manual to Accompany
The Many Worlds of Logic

By Paul Herrick
With illustrations by Steve Perlman

New York   Oxford
OXFORD UNIVERSITY PRESS

Oxford University Press

Oxford   New York
Auckland   Bangkok   Buenos Aires   Cape Town   Chennai
Dar es Salaam   Delhi   Hong Kong   Istanbul   Karachi   Kolkata
Kuala Lumpur   Madrid   Melbourne   Mexico City   Mumbai   Nairobi
São Paulo   Shanghai   Taipei   Tokyo   Toronto

Published by Oxford University Press, Inc.
198 Madison Avenue, New York, New York 10016
www.oup.com

Oxford is a registered trademark of Oxford University Press

ISBN: 0-19-515583-1

Printing number: 9 8 7 6 5 4 3 2 1

Printed in the United States of America
on acid-free paper

Preface

So, you're going to study logic. The study of logic may not sound as exciting as a class in the history of rock n' roll. A logic course may not be as immediately relevant as a class in personal finance or auto repair. But the study of logical theory can be every bit as useful and interesting. Logical theory has been considered a vital part of a university education for over 2000 years. And there is a reason for this. The study of logical theory helps train your mind. It makes you aware of thought-patterns you hadn't noticed before. It teaches correct forms of reasoning. It aids you in organizing and presenting your own thoughts. And it introduces you to an important part of the intellectual history of humankind.

After studying logical theory, you will be more skilled at understanding and evaluating the logical arguments that others present to you, and you will also be better at presenting good arguments of your own. Logic is a tool, and just as you can use a tool better when you understand its nature, you will reason more effectively when you acquire a deeper understanding of logical theory.

So, may your studies of logical theory be fruitful. I hope that at the end of the course you will feel that you have learned something worthwhile.

Answers to Exercises

The Selected Answers section at the back of the textbook, *The Many Worlds of Logic*, contains answers to approximately one third of the problems in the text. To save space, those answers are not repeated in this study guide. This study guide provides most of the rest of the answers. However, in each chapter, you will notice that a few problems have been left unanswered. Those problems were left unanswered so that they could be assigned by teachers as homework. So, problems that are not answered in this study guide and that are also not answered in the Selected Answers section of the textbook are problems that were purposely left unanswered so that they could be assigned as homework. On the homework problems, you're really on your own!

Table of Contents

Chapter 1

This opening chapter introduces several fundamental logical concepts, including the concepts of *argument*, *premise*, *conclusion*, *deduction*, *induction*, *validity*, *strength*, and *soundness*. After you finish this chapter, you should be able to:

- distinguish arguments from nonarguments
- distinguish premises from conclusions
- distinguish deductively valid arguments from deductively invalid arguments.
- distinguish inductively strong arguments from inductively weak arguments.
- distinguish between validity and soundness

Some comments

On Validity It is important to remember this: "valid" does not mean "true" (and "invalid" does not mean "false"). Validity concerns the *relation* between the truth-value of the premises and the truth-value of the conclusion. In a valid argument, the premises are related to the conclusion in such a manner that *if* the premises were to be true, the conclusion would have to be true as well, which is to say that it would be impossible for the premises to be true with the conclusion false. An argument is valid in virtue of this *relationship* (between premises and conclusion) and not in virtue of having actually true premises. Thus, in the case of a valid argument, it is *not* required that the premises be true.

On premises and conclusions. One way to understand something is to analyze it. ("Analysis" is from the Greek work *analytikos* meaning "to resolve into elements.") If you become skilled at distinguishing premises from conclusions and at identifying the various parts of an argument, and if you understand how the parts of an argument fit together, you will have a deeper understanding of argumentation and reasoning in general. As a result, your critical thinking skills will be enhanced. This means you will be better at understanding and evaluating the arguments others present, and you will be better at presenting good arguments of your own.

A Note on the Distinction Between Deductive and Inductive Arguments

It is important to remember this: the general distinction between deductive and inductive arguments has to do with the relationship that is *claimed*, by the arguer, to exist between the premises and the conclusion of an argument. Essentially, a deductive argument *claims* that the conclusion is related to the premises in such a way that *if* the premises were to be true, the conclusion would have to be true as well. In contrast, an inductive argument *claims* that the conclusion is related to the premises in such a way that *if* the premises were to be true, the conclusion would be probable.

If an author wishes to clarify his intentions and indicate that his argument is deductive, the arguer may introduce the conclusion with phrases such as:

It must be true that
It is clearly true that
I am absolutely certain that
There can be no doubt that
It is definite that
I am absolutely sure that
There is no other possibility
There can be no question that

For example:

1. All mammals have hair.
2. All aardvarks are mammals.
3. So, *it is definite that* all aardvarks have hair.

1. If the chemical is an acid, then the test paper will turn red.
2. The chemical is an acid.
3. Therefore, *there can be no doubt* that the test paper will turn red.

If an author wishes to explicitly state an *inductive* claim, she might introduce the conclusion with phrases such as

It is most likely that
It is most reasonable to suppose
The evidence is strong that
It is most likely that
We cannot be certain, but it seems that
The most reasonable conclusion is that
I'm fairly sure that
You can bet that

For example:

1. The concert is scheduled for August.
2. It hasn't rained in these parts in August for over 100 years.
3. So, I'm fairly sure that the concert won't be rained out.

1. Mr. Smeed, the Gottrocks's butler, was the last person to be seen in the vicinity of the jewels. 2. Before he left, the jewels were in their normal place (in the Gottrocks's closet). Immediately after he left, the jewels were found to be missing. 3. Mr. Smeed has mysteriously disappeared. 4. A background check revealed that he was convicted of burglary five times, and ten years ago he was a notorious jewel thief. 5. The *most reasonable conclusion* is that Smeed stole the jewels.

Answers

Exercise 1.1.

2. This is an explanation rather than an argument.
3. This is not an argument.
6. This is not an argument.
8. This is not an argument.
9. Argument. The conclusion is: "Maguire is going to lose the election."
11. This is not an argument.
13. Argument. The conclusion is: "Spiders are not insects."

Exercise 1.2

2. Inductive	8. Inductive
4. Inductive	10. Inductive
7. Deductive	12. Deductive

Exercise 1.3 Part I

2. Valid	8. Invalid
5. Invalid	12. Invalid
7. Invalid	13. Invalid.

Part II.
Deductively valid: 1,5,6,7,11,12*
Inductively strong: 2,3,4,8,9

*12 is valid if we assume Austin Powers is battling Dr. Evil.

Exercise 1.4

2. True	10. False
5. True	11. False
6. True	12. False

Exercise 1.5

2. Ann will swim today.
4. Each thing that exists has a cause.
6. Each existent entity has a location in space.
8. Spiders do not have six legs.
9. Every time Ed scowls, he is unhappy.
11. Las Vegas is located in Nevada.

Supplementary Exercises

Would you like some additional problems to practice? In case you would, here are some "supplementary" exercises. You might also use these as tests to make sure you are really understanding the concepts. Answers are provided after Supplementary Exercise 1.D, but don't peek at the answers until you've finished!

Supplementary Exercise 1.A

In each case, is the passage an argument or a nonargument?

1. Jane: You're a snob.
   Pete: You're a brat.
   Jane: No I'm not.
   Pete: Yes you are.
   Jane: No I'm not.
   Pete: Yes you are.

2. Well, we saw the herd of buffalo coming down the hill at about 8 A.M., but we were still in our pajamas and our cameras were still in the packing crates. So, we weren't able to take any photos.

3. The difference between us is this: You like tradition and I like novelty and change. So, that's why we don't always see eye to eye.

4. That chemical is a strong acid. A strong acid will harm you if you drink it. Therefore, that chemical will harm you if you drink it.

5. First, I had two Egg McMuffins. Then I had two breakfast burritos. Then I had a donut and a cheese danish. Then they started cooking hamburgers and so I had a quarter-pounder with cheese and some onion rings and some fries. I think that is why I'm feeling sick right now.

6. Fred has eaten two Big Macs for lunch every day for the past ten years, but he doesn't have heart problems. He also eats fish n' chips for dinner almost every night, except when he eats a couple bean and cheese burritos or some corndogs.

7. If the remaining Beatles have a reunion, then it will be one of the biggest events in the history of rock n' roll. Tens of Millions of people would want to attend the concert, especially aging baby-boomers who would hope to recapture a moment of their lost youth. Therefore, if the remaining Beatles ever hold a reunion concert, I'm sure it will be sold out.

Supplementary Exercise 1. B

Assume that the following arguments are all *deductive* arguments. In each case, is the argument valid or invalid?

1. Either we will eat fish or we will eat ham. But we won't eat fish. So, we will certainly eat ham.

2. Neither Ann nor Bob is home. So, Bob is surely not home.

3. If Ann wins the lottery, then Bob will be happy. Ann won't win the lottery. So Bob definitely won't be happy.

4. Ann is home. Bob is home. So, Ann and Bob must both be home.

5. If Ann goes swimming, then Bob will go swimming. If Bob goes swimming, then Rita will go swimming. So, certainly if Ann goes swimming, then Rita will go swimming.

6. No aardvarks are purple. No purple things are cute. So, necessarily, no aardvarks are cute.

7. Some illie pies are brown. Some brown things are heavy. So, some illie pies are heavy.

8. If Ann goes swimming, then Bob will go swimming. If Bob goes swimming, then Rita will go swimming. So, certainly if Ann does *not* go swimming, then Bob will *not* go swimming.

9. Either we will swim or we will jog. If we jog then we will miss dinner. So, necessarily, either we will swim or we will miss dinner.

10. We'll either eat burritos or hamburgers for dinner. If we eat burritos, then we'll eat nachos with dinner. If we eat hamburgers, then we'll eat fries with dinner. So, it is certain that we'll either eat nachos with dinner or fries with dinner.

11. All aardvarks are mammals. No mammals are purple. So, it certainly follows that no aardvarks are purple.

12. If we drive, then we'll go in Fred's car. If we walk, then we'll go down Third street. We'll go down Third street. So, it must be that we'll walk.

13. Jan will go swimming and Rita will go jogging. So, it necessarily follows that Jan will go swimming.

14. All aardvarks are mammals. Some mammals are cute. So, some aardvarks are cute.

Supplementary Exercise 1. C

Suppose the following arguments are *inductive*. In each case, is the argument strong or weak?

1. Every time that Pete has gone to Dick's Drive-in to eat, he's gotten a deluxe burger. He's eaten there twenty times this year. He just went to Dick's. It is very likely that he'll order a deluxe burger.

2. When Ann eats at Spud's Fish n' Chips, she sometimes orders extra fries. She is eating there tonight. She will probably order extra fries.

3. Professor Brown was late to class three times last week. He's rarely late for anything. He'll probably be late again this week.

4. Someone stole the Gottrocks's jewels. The only person seen near the jewels at the time was Joe the gardener. Therefore, we can be fairly sure that Joe did it.

5. Fred has owned eight Chevrolets and all eight were reliable. For his next car, he plans to buy another Chevrolet. His next car will likely be reliable.

6. When Kramer barges into Jerry's apartment, Kramer usually gets something to eat. Kramer just barged into Jerry's apartment. So, it is reasonable to conclude that he will get something to eat.

7. The remaining Beatles are not planning a reunion. Also, they do not seem to be re-establishing their relationships. So, we can be fairly sure that they won't be holding a reunion concert next month.

8. The remaining Beatles are not planning a reunion. Also, they do not seem to be re-establishing their relationships. So, it is very likely that they will never hold a reunion concert.

Supplementary Exercise 1. D

Each of the following is an *enthememe*. In each case, fill in the missing element (i.e., add a premise or conclusion) so as to turn the enthememe into a valid deductive argument.

1. All illie pies are green. Therefore, all illie pies are hairy.

2. No illie pies are heavy. Only heavy things are old. Therefore...

3. Some illie pies are green. So, some illie pies are young.

4. All illie pies are purple. No purple things are cute. So...

5. No green things are wrinkly. So, no illie pies are wrinkly.

6. If it rains then we'll stop the show. So, we will stop the show.

7. We will either eat burritos or gardenburgers. Consequently, we will eat gardenburgers.

8. Everybody loves Raymond. Therefore, Sue loves Raymond.

9. All aardvarks are mammals. Thus, no aardvarks are purple.

10. If it snows then we'll go skiing. It will snow. So...

Answers to Supplementary Exercises 1.A through 1.D

Exercise 1.A. Only 4 and 7 are arguments.

Exercise 1.B. Valid: 1, 2, 4, 5, 9, 10, 11, 13.

Exercise 1.C. Strong: 1, 5, 6, 7.

Exercise 1.D.

1. All green things are hairy.
2. No illie pies are old.
3. All green things are young.
4. No illie pies are cute.
5. All illie pies are green.

6. It will rain.
7. We won't eat burritos.
8. Sue is somebody.
9. No mammals are purple.
10. We'll go skiing.

Chapter 2

The main aim of this chapter is to introduce the five truth-functional operators. You should learn the *meaning* of each operator and not merely memorize the associated tables. (It will be much easier to work with the symbols if you understand the meaning of each one and have not merely memorized a set of abstract patterns.)

In addition to the five operators, this chapter introduces the following key logical concepts:

- The concept of a truth-functional sentence operator.
- The concept of a truth-functional compound sentence.
- The difference between a variable and a constant.
- the difference between simple and compound sentences

Answers

Exercise 2.1

2. Operator: "If, then"
Embedded sentences: "Bubba had....again" and "his wife...condition."

4. Operators: "It is not the case that," "and," "It is not the case that."
Embedded sentences: "Gilligan...caviar," "the Skipper...coconut."

7. Operator: "It is not the case that"
Embedded sentence: "Colonel Klink is vain."

9

8. Operator: "If, then," "and"
Embedded sentences: "the Blues Brothers...Rawhide," "those cowboys...bottles," "the band...alive."

9. Operators: "It is not the case that," "or," "It is not the case that"
Embedded sentences: "Kramer is hungry," "Jerry's...in it."

Exercise 2.2.

2. false.	8. true.
5. false.	9. true
6. true	10. false.

Chapter 3

The main aim of this chapter is to teach you how to translate information from English into truth-functional symbols. In the process, you will learn to *analyze* English sentences and abbreviate the parts with logical symbolism. In this chapter, you will also learn the logic pertaining to *necessary* and *sufficient conditions*, and you will learn to symbolize sentences reporting those conditions. The logic of necessary and sufficient conditions is an important part of modern scientific reasoning, as you will see in Chapter 26.

It is important to remember this: The truth-functional formulas represent only the *truth-functional* meaning of the corresponding English sentences. The formulas do not represent the entire meaning, which generally includes aspects of context, emotional overtones, nuances of various kinds, and more. In truth-functional logic, what matters is this: The truth-value of a formula will have a truth-functional relationship to the truth-values of the formula's components, and this relationship will be precisely the same truth-functional relationship the corresponding English sentence has to its sentential components.

Learning to translate English into logical symbols should enhance your understanding of language, for we generally understand something better when we resolve it into its elements and see how the elements fit together. Translation requires that you pay close attention to the logical structure of the English sentence; it requires that you distinguish operators from components; it has you determine which operator applies to which component; and it requires that you correlate English grammatical devices with logical structure. I believe that this deepens your understanding of language and reasoning. You might say that in this chapter we use logic to "dissect" sentences.

Here is another point. Computers are now everywhere. Working with computers requires precise and systematic thinking. One small error (such as hitting a sequence of keys in the wrong order) can have large repercussions resulting in a major error in the end. Likewise in a logic problem, one small mistake in translation can result in a large error when evaluating a sentence or argument. Symbolic logic thus gives you practical experience thinking precisely and systematically, and this type of thinking is used in many important fields. (Similarly, in math problems, a small error made at the beginning of a calculation can throw off the rest of the problem, resulting in a major error at the end.)

Answers

Exercise 3.1

2. A & (B v D)

5. D & (C v O)

6. J & (B & G)

9. ~(G & J) or: ~G v ~J

10. ~L & M

12. ~M v ~S

13. (C v M) v (Y & B)

16. (~H & ~S) & (V v L)

17. (M & D) & G

Exercise 3.2

2. (B & L) ⊃ C

5. ~B ⊃ (C v D)

6. ~L ⊃ (B v G)

9. (E & P) ≡ (A & ~T)

10. (E & R) ⊃ (B & D) 20. ~B ⊃ J

12. S ⊃ C

15. A ≡ (B & S)

16. G ⊃ R

18. ~(G ⊃ M)

12

Chapter 4

The main aim of this chapter is to teach how to calculate the truth-value of a truth-functional compound sentence. In this chapter, you will also learn the syntax of the language TL. This chapter is a prerequisite for the truth-tables of Chapters 5 and 6.

Answers

Exercise 4.1

Wffs: 1,2,5,7,8,13,16

Exercise 4.2

2. second wedge from left   8. the horseshoe
5. the ampersand   9. the triple bar
6. third wedge

Exercise 4.3

2. F	10. T
3. F	13. F
6. T	14. T
8. T	17. F
	19. T

Exercise 4.4

2. F	10. T
3. F	12. F
6. T	14. F
8. F	15. T

Exercise 4.5

2. T	13. T
5. T	14. T
6. T	17. T
9. T	18. T
10. T	

Exercise 4.6

3. T	10. F
5. T	12. F
8. T	14. T
9. T	15. T

Chapter 5

In this chapter, you will use *truth-tables* to precisely determine

- whether a sentence is tautologous, contradictory, or contingent.
- whether an argument is valid or invalid.
- whether two sentences are equivalent or not.

Answers

Exercise 5.1

```
2. A| ~( A ⊃ A)
    T| F   T T T
    F| F   F T F  (Contradiction)
```

```
5. A B E| (A v B) ⊃ (A & E)
   T T T|  T T T  T  T T T
   T T F|  T T T  F  T F F
   T F T|  T T F  T  T T T
   T F F|  T T F  F  T F F
   F T T|  F T T  F  F F T
   F T F|  F T T  F  F F F
   F F T|  F F F  T  F F T
   F F F|  F F F  T  F F F  (Contingent)
```

```
6. A B| A ⊃ (A v B)
   T T| T T  T T T
   T F| T T  T T F
   F T| F T  F T T
   F F| F T  F F F  (Tautology)
```

```
12. A| A ≡ ~A
    T| T F FT
    F| F F TF  (Contradiction)
```

```
13. A| A ≡ (A v A)
    T| T T  T T T
    F| F T  F F F  (Tautology)
```

```
14. A B| (A & B) ⊃ B
    T T|  T T T  T T
    T F|  T F F  T F
    F T|  F F T  T T
    F F|  F F F  T F  (Tautology)
```

14

15. A B | (A & B) ⊃ (~A ⊃ ~B)

A B	(A & B)	⊃	(~A ⊃ ~B)	
T T	T T T	T	FT T FT	
T F	T F F	T	FT T TF	
F T	F F T	T	TF F FT	
F F	F F F	T	TF T TF	(Tautology)

18. A B | [(A v B) & ~A] ⊃ B

A B	[(A v B) & ~A]	⊃	B	
T T	T T T F FT	T	T	
T F	T T F F FT	T	F	
F T	F T T T TF	T	T	
F F	F F F F TF	T	F	(Tautology)

19. A B | (A & ~A)⊃ B

A B	(A & ~A)	⊃	B	
T T	T F FT	T	T	
T F	T F FT	T	F	
F T	F F TF	T	T	
F F	F F TF	T	F	(Tautology)

21. A B | (A ⊃ B) ≡ (~A v B)

A B	(A ⊃ B)	≡	(~A v B)	
T T	T T T	T	FT T T	
T F	T F F	T	FT F F	
F T	F T T	T	TF T T	
F F	F T F	T	TF T F	(Tautology)

22. A B | (A ⊃ B) ≡ (~B ⊃ ~A)

A B	(A ⊃ B)	≡	(~B ⊃ ~A)	
T T	T T T	T	FT T FT	
T F	T F F	T	TF F FT	
F T	F T T	T	FT T TF	
F F	F T F	T	TF T TF	(Tautology)

23. A B | (A & B) ⊃ ~(~A v ~B)

A B	(A & B)	⊃	~(~A v ~B)	
T T	T T T	T	T FT F FT	
T F	T F F	T	F FT T TF	
F T	F F T	T	F TF T FT	
F F	F F F	T	F TF T TF	(Tautology)

26. A B | A ⊃ (~B ⊃ A)

A B	A	⊃	(~B ⊃ A)	
T T	T	T	FT T T	
T F	T	T	TF T T	
F T	F	T	FT T F	
F F	F	T	TF F F	(Tautology)

27. A B | ~A ⊃ (A ⊃ B)

A B	~A	⊃	(A ⊃ B)	
T T	FT	T	T T T	
T F	FT	T	T F F	
F T	TF	T	F T T	
F F	TF	T	F T F	(Tautology)

15

```
29. A | ~( A & ~A)
    T |  T   T F FT
    F |  T   F F TF  (Tautology)
```

```
31. A B |(A & ~A)  v  (B & ~B)
    T T | T F FT    F   T F FT
    T F | T F FT    F   F F TF
    F T | F F TF    F   T F FT
    F F | F F TF    F   F F TF  (Contradiction)
```

Exercise 5.2

```
2.  A B |(A v B)    ~A      / ~B
    T T | T T T     FT        FT
    T F | T T F     FT        TF
    F T | F T T     TF        FT     ←
    F F | F F F     TF        TF   Invalid
```

```
4.  A B |(A ⊃ B)   / ~A ⊃ ~B
    T T | T T T      FT T FT
    T F | T F F      FT T TF
    F T | F T T      TF F FT    ←
    F F | F T F      TF T TF  Invalid
```

```
6.  Invalid
```

```
8.  A B | A ⊃ B  / ~B ⊃ ~A
    T T | T T T    FT T FT
    T F | T F F    TF F FT
    F T | F T T    FT T TF
    F F | F T F    TF T TF   Valid
```

```
9.  A B | A ⊃ B    ~B / ~A
    T T | T T T     FT    FT
    T F | T F F     TF    FT
    F T | F T T     FT    TF
    F F | F T F     TF    TF   Valid
```

```
10. A B | A ⊃ B    A / B
    T T | T T T     T   T
    T F | T F F     T   F
    F T | F T T     F   T
    F F | F T F     F   F   Valid
```

16

13.

ABC	~[A & (B & C)] / ~A & (~B & ~C)
TTT	F T T T T T FT F FT F FT
TTF	T T F T F F FT F FT F TF ←
TFT	
TFF	
FTT	
FTF	
FFT	
FFF	

Invalid

14.

A B	~A v ~B / ~(A & B)
T T	FT F FT F T T T
T F	FT T TF T T F F
F T	TF T FT T F F T
F F	TF T TF T F F F Valid

15.

A B	~(A v B) / ~A & ~B)
T T	F T T T F T F FT
T F	F T T F F T F TF
F T	F F T T T F F FT
F F	T F F F T F T TF Valid

Exercise 5.3

2.
```
A ⊃ B   ~A/~A & ~B
F T T   TFTF F FT
```

3.
```
P ⊃ (~Q & ~R)   (~R ⊃ S) & ~D /S ⊃ P
FT   FT F FT     FT TT   T TF T F F
```

4.
```
J ⊃ B   ~J /~B
FTT     TF FT
```

7.
```
(~A & ~B)&(C & D)   (A v C) ⊃ R /R ⊃ A
 TF T TF T T T T     F T T T T  T F F
```

8.
```
~A&~B   ~B⊃(CvD)   ~R⊃(~C&~D)   /~R v C
TFTTF   TFT FTT    FTT TFFFT     FTFF
```

Exercise 5.4

2.
```
A ⊃ B   W ⊃ S   ~A v ~W   /~B v ~S
TTT     FTT     FT TTF     FT F FT
```

5.
```
A ⊃ B   B ⊃ W   W ⊃ S   S ⊃ H   / H ⊃ A
FTT     TTT     TTT     TTT       TFF
```

17

6. $\underline{A \supset B \ /B \supset A}$
    FTT   TFF

7. $\underline{\sim(A \ \& \ B)/\sim A \ \& \ \sim B}$
    T TFF   FT F TF

8. $\underline{\sim(A \ v \ B) \qquad / \ \sim(\sim A \ v \ \sim B)}$
       T F F F   F TF T TF

11. $\underline{A \supset B \quad W \supset B \ / \ A \supset W}$
    T T T   FT T T F F

13. $\underline{A \supset B \quad B \ /A}$
    F T T   T F

14. $\underline{A \supset B \quad \sim A \quad /\sim B}$
    F T T   TF   FT

Exercise 5.5

2. $\underline{P \supset R \quad \sim P \quad /\sim R}$
    F T T   TF   FT   Invalid

4.

GM	$\sim$(G & M)	M	/	$\sim$G
TT	F	T		T	F
TF	T	F		F	F
FT	T	F		T	T
FF	T	F		F	T	Valid

4.

GM	~(G & M)	M /	~G	
TT	F T	T	F	
TF	T F	F	F	
FT	T F	T	T	
FF	T F	F	T	Valid

6. 1. $\sim(K \ \& \ S)$
    2. $\sim K \supset M$
    3. $\underline{\sim S \supset M} \ \ / \ M$ Valid

8. $\underline{(RvA) \supset N/\sim Rv\sim N}$
                           Invalid
      TTF TT FTFFT

9. $\underline{M \supset (L\&C) \quad \sim M \ /\sim Lv\sim C}$
    FT TTT   TF FTFFT Invalid

10.

SWC	(S & ~W) v (S & C)	S & ~C /	~W	
TTT	T	F	F	
TTF	F	T	F	
TFT	T	F	T	
TFF	T	T	T	
FTT	F	F	F	
FTF	F	F	F	
FFT	F	F	T	
FFF	F	F	T	Valid

11.

M ⊃ (C ⊃ L)	C ⊃ (M ⊃ L)	~C & ~M /~L	
F T F T T	F T F T T	TF T TF FT	Invalid

Exercise 5.6

2.

P	~~P	P
T	T	T
F	F	F

Equivalent

3.

P Q	~(P & Q)	~ P v ~Q
T T	F	F
T F	T	T
F T	T	T
F F	T	T

Equivalent

6.

P Q	P v Q	~(~P & ~Q)
T T	T	T
T F	T	T
F T	T	T
F F	F	F

Equivalent

8.

PQR	P & (Q v R)	(P & Q) v (P & R)
TTT	T	T
TTF	T	T
TFT	T	T
TFF	F	F
FTT	F	F
FTF	F	F
FFT	F	F
FFF	F	F

Equivalent

10.

P	Q	P⊃Q	~Q⊃~P
T	T	T	T
T	F	F	F
F	T	T	T
F	F	T	T

Equivalent

12.

P	Q	~(~P v Q)	~(P & ~Q)
T	T	F	T
T	F	T	F
F	T	F	T
F	F	F	T

Not equivalent

13.

P	Q	~(P ⊃ Q)	(~P ⊃ ~Q)
T	T	F	T
T	F	T	T
F	T	F	F
F	F	F	T

Not Equivalent

15.

P	P v P	P
T	T	T
F	F	F

Equivalent

16.

P	Q	P & Q	Q & P
T	T	T	T
T	F	F	F
F	T	F	F
F	F	F	F

Equivalent

18. Equivalent
19. Equivalent
21. Equivalent
23. Not equivalent
24. Equivalent
25. Not equivalent

Chapter 6

The main aim of this chapter is to explain, in precise terms:

- the concept of a sentence form
- the concept of an argument form.
- the concept of a *valid* argument form.

This chapter lays the groundwork for the natural deduction system of Chapter 7. Before you begin Chapter 7, it is crucial that you understand the following:

1. The difference between a constant and a variable.
2. The difference between a form and its instances.
3. The difference between a valid and an invalid argument form.

Answers

Exercise 6.1

c: 1, 8, 9, 15, 16   j: 3,13,18
e: 8,15   k: 13
f: 6   l: 12
h: 6, 14   o: 5,10,17

Exercise 6.2

3. Invalid   15. Invalid
4. Invalid   17. Valid
6. Valid   18. Valid
9. Valid   19. Valid
10. Invalid   20. Valid
13. Invalid   23. Valid

Exercise 6.3

1. Tautology. (The sentence is an instance of a tautological form.)

3. Tautology. (The sentence is an instance of a tautological form.)

Supplementary Exercise 6.A

In case you would like more practice identifying forms, here is a supplementary exercise. Answers are provided below.

Which of the following symbolized arguments are instances of:

1. The Modus Ponens form

2. The Modus Tollens form

3. The Disjunctive Syllogism form.

4. The Hypothetical Syllogism form

1. $H \supset E$	2. $H \supset E$	3. $H \supset E$	4. $H \supset E$	5. $H \supset S$	6. $H \supset E$
$E \supset C$	~E	$E \supset C$	~H	S	H
$H \supset C$	~H	$C \supset H$	~E	H	E

7. $R \supset P$	8. $H \supset E$	9. ~$H \supset$ ~S	10. $(H \& E) \supset S$	11. ~Hv~S
$G \supset S$	$E \supset I$	~H	~S	~~H
PvS	HvE	~S	~(H&E)	~S
RvG	EvI			

22

12. (H&E)⊃(FvE)
    (H&E)
    ―――――
    (FvE)

13. (EvS)⊃(U&T)
    ~(U&T)
    ―――――
    ~(EvS)

14. Hv~S
    ~H
    ――
    ~S

15. HvE
    H
    ――
    E

16. (H&E)v~M
    ~H&E
    ―――
    ~M

17. (H&E)⊃E
    E⊃(RvS)
    ―――――
    (H&E)⊃(RvS)

18. ~Gv~R
    ~G
    ――
    ~R

19. (R&H)⊃S
    S
    ――
    R&H

20. (U⊃S)⊃~H
    ~~H
    ―――――
    ~(U⊃S)

21. HvE
    EvC
    ――
    HvC

Answers

The Disjunctive Syllogism form: 11, 14
The Hypothetical Syllogism form: 1, 17
The Modus Ponens form: 6, 9, 12
The Modus Tollens form: 2, 10, 13, 20

Chapter 7

In this chapter, you will learn to use a natural deduction system to *prove* arguments valid. What's the value of natural deduction? First, using the inference rules of a natural deduction system, we can take an extremely complicated valid argument and definitively prove it valid, in very precise terms, whereas if we tried to reason it out in our heads, "intuitively," it would be nearly impossible to reach a definitive judgment. Second, computers use systems akin to natural deduction to "reason" their way through various types of problems, and so an understanding of natural deduction introduces you to the nature of the computational processes that go on inside a digital computer. Third, working with natural deduction proofs helps develop your ability to recognize abstract patterns, and pattern recognition is an important mental skill.

When we translate English into the symbols of our formal language TL, the formal language clearly displays the parts of the sentences and the logical relations among the parts. We then apply formal operations to the symbols to produce a definite answer to a logical problem. The answer is reached in a precise, systematic way. Working with a formal system thus teaches us something about precise and systematic thinking. For instance, information in English has to be translated precisely into symbols because a small error made at the beginning of a problem can throw the entire problem off track; rules must be applied accurately or incorrect proofs will result, and so on.

Finally, you will notice that in natural deduction problems, our assumptions and premises are stated explicitly and precisely, "up front." This is something we should strive for when we reason with others. In these respects, natural deduction can serve as a model for clear and precise thought.

Answers

Exercise 7.1

2.	6.	DS	3,4
	7.	MP	2,6
	8.	DS	1,7
	9.	MP	5,8
3.	5.	MT	1,3
	6.	MP	2,5
	7.	MP	4,6
6.	5.	DS	3,4
	6.	MT	1,5
8.	7.	MP	1,5
	8.	MP	2,6
	9.	DS	7,8
	10.	MP	3,9
	11.	HS	4,10
9.	6.	MP	1,2

```
      7.   DS   3,6
      8.   MP   5,7
      9.   MT   4,8

12.   5.   MP   1,3
      6.   DS   2,4
      7.   HS   5,6

14.   4.   DS   1,2
      5.   MP   3,4
      6.   MP   2,5
      7.   MP   4,6

19.   5.   HS   1,2
      6.   HS   3,5

20.   5.  HS 3,4
      6.  MP 1,5
      7.  HS 4,6
      8.  HS 3,7
      9.  MP 2,8
     10.  MP 7,9
```

Exercise 7.2

```
(2)1.GvR
2.H&(J⊃I)
3.~G   /R
4.  R            DS  1,3

(3)1.~(F≡S)
2. RvM
3.~(F≡S)⊃D / D
4. D             MP  1,3

(4)1.~(A&B)
2.R⊃(A&B)
3.H /~R
4.  ~R            MT  1,2

(6)1.(B&C)v A
2.A⊃F
3.~(B&C) /F
4. A            DS  1,3
5. F            MP  2,4
```

(8) 1. (E≡F) v ~(A&B)
2. H⊃(A&B)
3. ~(E≡F) / ~ H
4. ~(A&B)   DS   1,3
5. ~H   MT   2,4

(9) 1. F⊃S
2. S⊃G
3. (F⊃G)⊃M   / M
4. F⊃G   HS   1,2
5. M   MP   3,4

(12) 1. B⊃(A&G)
2. R⊃B
3. [R⊃(A&G)]⊃S / S
4. R⊃(A&G)   HS   1,2
5. S   MP   3,4

(14) 1. J⊃I
2. I⊃~R
3. J
4. A⊃~B
5. A
6. Rv(BvS).
7. S⊃L / L
8. J⊃ ~R   HS   1,2
9. ~R   MP   3,8
10. BvS   DS   6,9
11. ~B   MP   4,5
12. S   DS   10,11
13. L   MP   7,12

Exercise 7.3

(2) 1. A ⊃ B
2. B ⊃ C
3. ~C   /~A
4. A ⊃ C   HS   1,2
5. ~A   MT   3,4

(4) 1. (PvQ)⊃(R&W)
 2. H ⊃(PvQ)   /H ⊃(R&W)
 3. H ⊃(R&W)   HS   1,2

(6)1.T ⊃(A ⊃I)
 2.~I
 3.~S ⊃ T
 4.Iv~S___ /~A
 5. ~S   DS   2,4
 6. T   MP   3,5
 7. A ⊃ I   MP   1,6
 8. ~A   MT   2,7

(9)1.(A ⊃B)⊃C
 2.A v ~D
 3.~D ⊃(A ⊃ B)
 4.~A_/C
 5. ~D   DS   2,4
 6. A ⊃ B   MP   3,5
 7. C   MP   1,6

(10)1.~(A&B)
 2.(CvR)v M
 3.~L ⊃~(CvR)
 4.(A&B)v ~L___ /M
 5. ~L   DS   1,4
 6. ~(CvR)   MP   3,5
 7. M   DS   2,6

(11)1.W ⊃ P
 2.Ov~P
 3.~O___ / ~W
 4. ~P   DS   2,3
 5. ~W   MT   1,4

(14)1.Hv(S ⊃B)
2.(S ⊃B) ⊃(B ⊃T)
3.~H
4.~T___/~S
5. S ⊃B   DS   1,3
6. B ⊃T   MP   2,5
7. S ⊃T   HS   5,6
8. ~S   MT   4,7

(15)1.G ⊃ S
2.S ⊃ ~E
3.~E ⊃ M
4.(G ⊃M)⊃J___ / J   6. G ⊃ M   HS   5,3
5. G ⊃ ~E   HS   1,2   7. J   MP   4,6

27

Chapter 8

When the four rules of this chapter are added to the four rules of the previous chapter, many valid truth-functional arguments can be *proven* valid. The eight rules produce a powerful natural deduction system.

Note the list of common deduction errors at the end of Chapter 8 in the textbook. It may be helpful to go through these very carefully and see *why* each is an error.

Answers

Exercise 8.1

```
2.   6.   Conj  2,3
     7.   MP    4,6
     8.   MT    1,7
     9.   MP    5,8
    10.   Add   9

3.   5.   Add   4
     6.   MP    1,5
     7.   CD    2,3,6

4.   6.   Conj  1,2
     7.   MP    3,6
     8.   MT    4,7
     9.   Add   8
    10.   MP    5,9
    11.   Add   10
```

Exercise 8.2

```
(2)1. (AvB)⊃G
2. A
3. G⊃S    / S
4. AvB              Add   2
5. G                MP    1,4
6. S                MP    3,5

(3)1. (H&S)⊃~(F≡S)
2. B⊃(F≡S)
3. H
4. S /~B
5. H&S              Conj  3,4
6. ~(F≡S)           MP    1,5
7. ~B               MT    2,6
```

(6) 1. H⊃~(S v G)
2. ~(RvW)⊃(SvG)
3. F&H
4. ~W⊃(J&A)
5. W⊃M
6. ~M   / A& ~~(RvW)
7. H   Simp 3
8. ~(SvG)   MP 1,7
9. ~~(RvW)   MT 2,8
10. ~W   MT 5,6
11. J&A   MP 4,10
12. A   Simp 11
13. A & ~~(RvW)Conj 12,9

(8) 1. ~F⊃~S
2. H&F
3. F⊃B
4. B⊃G   / G
5. F   Simp 2
6. B   MP 5,3
7. G   MP 6,4

(11) 1. (HvS)⊃D
2. (RvM)⊃I
3. (D&I)⊃~A
4. H&R   /~A
5. H   Simp 4
6. R   Simp 4
7. HvS   Add 5
8. D   MP 1,7
9. RvM   Add 6
10. I   MP 2,9
11. D&I   Conj 8,10
12. ~A   MP 3,11

(12) 1. Av(B⊃C)
2. ~A⊃(~H⊃J)
3. ~HvB
4. ~A&I   /JvC
5. ~A   Simp 4
6. ~H ⊃ J   MP 2,5
7. B ⊃ C   DS 1,5
8. JvC   CD 3,6,7

(14) 1. (S⊃I)&(S⊃J)
2. (IvJ)⊃G
3. H&S   /G

29

```
4. S ⊃ I          Simp  1
5. S ⊃ J          Simp  1
6. S              Simp  3
7. SvS            Add   6
8. IvJ            CD    4,5,7
9. G              MP    2,8


(16) (A⊃J)&X
2.Fv(J⊃F)
3. ~F /~AvZ
4. J⊃ F           DS    2,3
5. A⊃ J           Simp  1
6. A⊃ F           HS    4,5
7. ~A             MT    3,6
8. ~AvZ           Add   7


(19)1.(J⊃I)&(~J⊃S)
2. I⊃B
3. [(J⊃I)&(I⊃B)]⊃[(J&B)v(~J&~B)]
4. (J&B)⊃A
5. (~J&~B)⊃H / AvH
6. J⊃ I                    Simp  1
7. (J⊃ I)&(I ⊃ B)          Conj  2,6
8. (J&B)v(~J&~B)           MP    3,7
9. AvH                     CD    4,5,8


(21)1.A⊃B
2. ~B&I
3. ~Sv~G
4. (~A&~B)⊃[(~S⊃A)&(~G⊃X)]   /X
5. ~B                      Simp  2
6. ~A                      MT    1,5
7. ~A&~B                   Conj  5,6
8. (~S ⊃A)&(~G ⊃X)         MP    4,7
9. ~S ⊃A                   Simp  8
10. ~G ⊃X                  Simp  8
11. AvX                    CD    3,9,10
12. X                      DS    6,11


(23)1.(AvB)⊃[(E⊃G)&(J⊃I)]
2. (AvE)⊃(EvJ)
3. A&B     /GvI
4. A                       Simp  3
5. AvB                     Add   4
6. (E⊃G)&(J⊃I)             MP    1,5
```

30

```
7.  AvE                  Add   4
8.  EvJ                  MP    2,7
9.  E⊃G                  Simp  6
10. J⊃I                  Simp  6
11. GvI                  CD    9,10,8

(24)1.~A&B
2.  ~A&(Jv~I)
3.  (J⊃S)&(~I⊃~H)
4.  (S⊃A) & (~H⊃~Z)   /~Z
5.  ~A                   Simp  1
6.  Jv~I                 Simp  2
7.  J⊃S                  Simp  3
8.  ~I⊃~H               Simp  3
9.  Sv~H                 CD    6,7,8
10. S⊃A                  Simp  4
11. ~H⊃~Z               Simp  4
12. Av~Z                 CD    9,10,11
13. ~Z                   DS    5,12
```

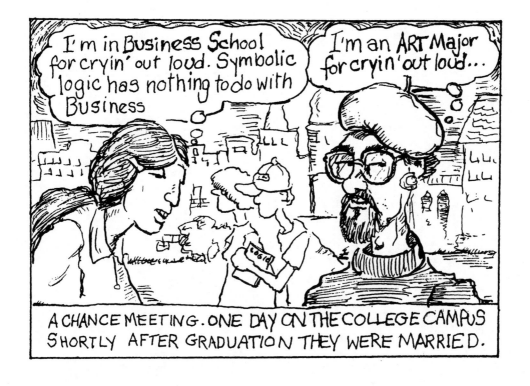

A CHANCE MEETING. ONE DAY ON THE COLLEGE CAMPUS SHORTLY AFTER GRADUATION THEY WERE MARRIED.

31

Chapter 9

This chapter introduces the methods of conditional and indirect proof. Using indirect and conditional proof, we can prove many arguments valid that could not otherwise have been proven valid. When conditional and indirect proof are added to the eight rules of the previous two chapters, the resulting natural deduction system is an even more powerful logical tool.

Answers

Exercise 9.1

```
(2)  1. (HvB)⊃(A&E)
2.  (EvS)⊃(G&~H)
3.  ~~H⊃H  /~H
4.              | ~~H           AP
5.              | H             MP 3,4
6.              | H v B         Add 5
7.              | A & E         MP 1,6
8.              | E             Simp 7
9.              | E v S         Add 8
10.             | G & ~H        MP 2,9
11.             | ~H            Simp 10
12.             | H & ~H        Conj 5, 11
13.    ~H                       IP 4-12
```

```
(3) 1. (~JvI)⊃(A&B)
2. (AvZ)⊃(B⊃J) / J
3.             | ~J     AP
4.             | ~J v I    Add 3
5.             | A & B     MP 1,4
6.             | A         Simp 5
7.             | A v Z     Add 6
8.             | B ⊃ J     MP 2,7
9.             | B         Simp 5
10.            | J         MP 8,9
11.            | J & ~J    Conj 3,10
12.   J           IP 3-11
```

```
(6) 1. G⊃N
2  ~A
3. ~G⊃A  / N
4.             | ~N   AP
5.             | ~G   MT 1,4
6.             | A    MP 3, 5
7.             | A & ~A Conj 2,6
8 N               IP 4-7
```

(7) 1. ~~M⊃(N⊃O)
 2. ~O⊃(M⊃N)
 3. ~M⊃O   / O
 4. | ~O   AP
 5. | ~~M   MT 3,4
 6. | N⊃O   MP 1,5
 7. | ~N   MT 4, 6
 8. | M⊃N   MP 2,4
 9. | ~M   MT 7,8
10. | ~M &~~M   Conj 5,9
11. O   IP 4-10

(8) 1. A⊃B
 2. ~B   /~ A
 3. | ~~A   AP
 4. | ~A   MT 1,2
 5. | ~A & ~~A   Conj 3,4
 6. ~A   IP 3-5

(11) 1. Av~(B&C)
 2. (B&C)v W
 3. ~(AvW)⊃(~A&~W)   /Av W
 4. | ~(AvW)   AP
 5. | ~A&~W   MP 3,4
 6. | ~A   Simp 5
 7. | ~(B&C)   DS 1,6
 8. | W   DS 2,7
 9. | ~W   Simp 5
10. | W & ~W   Conj 9,8
11. AvW   IP 4-10

(12) 1. A v ~~B
 2. ~BvA   / A
 3. | ~A   AP
 4. | ~~B   DS 1,3
 5. | A   DS 2,4
 6. | A&~A   Conj 3,5
 7.   A   IP 3-6

(13) 1. (AvE)⊃(B&H)
 2. ~B
 3. ~~A⊃A /~A
 4. | ~~A   AP
 5. | A   MP 3,4
 6. | AvE   Add 5
 7. | B&H   MP 1,6
 8. | B   Simp 7
 9. | B&~B   Conj 2,8

33

10.   ~A   IP 4-9

Exercise 9.1 **Part II**

(1) 1. (A&B)⊃S
2. H⊃R
3. (A&B) / S
4. | ~S   AP
5. | ~(A & B)   MT 1,4
6. | (A & B) & ~(A & B)   Conj 3, 5
7.   S   IP 4-6

(10) 1. R⊃H
2. H⊃S
3. S⊃G
4. (R⊃G)⊃F / F
5. | ~F   AP
6. | ~(R ⊃ G)   MT 4,5
7. | R ⊃ S   HS 1, 2
8. | R⊃G   HS 7, 3
9. | (R ⊃ G) & ~(R ⊃G)   Conj 8, 6
10   F   IP 5-9

Exercise 9.2

(2) 1. (IvI)⊃I /I⊃I
2. | I   AP
3. | IvI Add 2
4. | I   MP 1,3
5. I ⊃ I   CP 2-4

(4) 1. (A&B)⊃C
2. A⊃B   /A⊃C
3. | A   AP
4. | B   MP 2,3
5. | A&B   Conj 3,4
6. | C   MP 1,5
7.   A ⊃C   CP 3-6

(7) 1. A⊃~B
2. Bv(H&R) / A⊃(RvZ)
3. | A   AP
4. | ~B   MP 1,3
5. | H & R   DS 2,4

34

```
6.              |R                     Simp 5
7.              |R v Z                 Add 6
8. A⊃(R v Z)    CP 3-7


(9) 1. H⊃(S&~L)
2. Lv~~I
3. X⊃~I
4. ~X⊃B        /H⊃B
5.              |H                     AP
6.              |S&~L                  MP  1,5
7.              |~L                    Simp  6
8.              |~~I                   DS  2,7
9.              |~X                    MT  3,8
10.             |B                     MP  4,9
11.   H⊃B        CP  5-10


(10) 1. A⊃J
2. A⊃(J⊃B)
3. J⊃(B⊃T)   / A⊃T
4.              |A                     AP
5.              |J                     MP  1,4
6.              |J⊃B                   MP  2,4
7.              |B                     MP  5,6
8.              |B⊃T                   MP  3,5
9.              |T                     MP  7,8
10.   A⊃T        CP  4-9


(12) 1. A⊃B   /A⊃[(B v S) v G]
2.              |A                     AP
3.              |B                     MP 1,2
4.              |B v S                 Add 3
5.              |(B v S) v G           Add 4
6. A ⊃[(B v S) v G]                    CP 2-5

(14) 1. (E v G) ⊃ (Z & H)
2. (Z v M) ⊃ W      / E ⊃ W
3               |E                     AP
4.              |E v G                 Add 3
5.              |Z & H                 MP 1,4
6.              |Z                     Simp 5
7.              |Z v M                 Add 6
8.              |W                     MP 2,7
9.   E ⊃ W       CP 3-8


(15) 1. A ⊃ B
2. B ⊃ R    / A ⊃ R
```

```
3.                  |A              AP
4.                  |B              MP 1,3
5.                  |R              MP 2,4
6.   A ⊃ R        CP 3-5


(17) 1.E ⊃ M
2. E ⊃ W      / E ⊃ (M & W)
3.                  |E              AP
4.                  |M              MP 1,3
5.                  |W              MP 2,3
6.                  |M & W          Conj 4,5
7.   E ⊃(M & W) CP 3-6


(18) 1. B ⊃ D
2. C ⊃ E   / (B & C) ⊃ (D & E)
3.                  |B & C          AP
4                   |B              Simp 3
5.                  |C              Simp 3
6.                  |D              MP 1,4
7.                  |E              MP 2,5
8.                  |D & E          Conj 6,7
9.  (B & C)⊃(D & E)              CP 3-8


(20) 1. A ⊃ (B & C)
2. C ⊃ G
3. G ⊃ N      / A ⊃ (N v X)
4.              |A          AP
5.              |B & C      MP 1,4
6.              |C          Simp 5
7.              |G          MP 2,6
8.              |N          MP 3,7
9.              |N v X      Add 8
10.   A ⊃(N v X)          CP 4-9


(21) 1. (A & B) ⊃ (C & D)
2. B & ( C ⊃ ~ H)    / A ⊃ ~H
3.    |A          AP
4.    |B          Simp 2
5.    |A & B      Conj 3,4
6.    |C & D      MP 1,5
7.    |C          Simp 6
8.    |C ⊃~H      Simp 2
9.    |~H         MP 7,8
10.   A ⊃ ~H     CP 3-9
```

Exercise. 9.3

```
(2)1.A⊃(B⊃C)
2.  (CvH)⊃K  /A⊃(B⊃K)
3.       │A   AP
4.                │B   AP
5.                │B⊃C  MP  1,3
6.                │C   MP  4,5
7.                │CvH  Add  6
8.                │K   MP  2,7
9.       │B⊃K      CP  4-8
10. A⊃(B⊃K)   CP  3-9
```

```
(3)1.A⊃(B&E) /(J⊃A)⊃(J⊃E)
2.       │J⊃A  AP
3.                │J   AP
4.                │A   MP  2,3
5.                │B&E  MP  1,4
6.                │E  Simp  5
7.       │J⊃E  CP  3-6
8.(J⊃A)⊃(J⊃E)   CP  2-7
```

```
(6)1.A⊃B
2.  (A&B)⊃I
3.  (H&I)⊃S  /A⊃(H⊃S)
4.       │A        │AP
5.       │B        │MP  1,4
6.                 │H            AP
7.                 │A&B          Conj  4,5
8.                 │I            MP  2,7
9.                 │H&I          Conj  6,8
10.                │S            MP  9,3
11.      │ H⊃S  CP  6-10
12.   A⊃(H⊃S)  CP  4-11
```

```
(7) 1.(AvB)⊃[(CvD)⊃(E&H)]
2.  (E&H)⊃J  /A⊃(C⊃J)
3.  │A            │AP
4.                │C                      AP
5.                │A v B                  Add 3
6.                │(C v D) ⊃ (E & H)      MP 1,5
7.                │C v D                  Add 4
8.                │E & H                  MP 6,7
9.                │J                      MP 2,8
10. │C ⊃ J                            CP 4-9
```

11. A ⊃ (C ⊃ J)   CP 3-10

(8) 1. A⊃[W⊃(E&H)]
2. [A&(W&H)]⊃(D&~M) /A⊃(W⊃~M)
3.   │A   AP
4.   │ │W   AP
5.   │ │W ⊃(E&H)   MP 1,3
6.   │ │E&H   MP 4,5
7.   │ │H   Simp 6
8.   │ │W&H   Conj 4, 7
9.   │ │A&(W&H)   Conj 3, 8
10.   │ │D&~M   MP 2, 9
11.   │ │~M   Simp 10
12.   │W ⊃~M   CP 4-11
13.   A ⊃(W ⊃~M)   CP 3-12

Exercise 9.4

(2) 1.   │(J⊃I)&(I⊃I)]   AP
2.   │J⊃I   Simp 1
3.   │I⊃I   Simp 1
4.   │J⊃I   HS 2,3
5. [(J⊃I)&(I⊃I)]⊃(J⊃I)   CP 1-4

(3) 1.   │A   AP
2.   │ │A⊃B   AP
3.   │ │B   MP 1, 2
4.   │[(A⊃B)⊃B]   CP 2-3
5.   A⊃[(A⊃B)⊃B]   CP 1-4

(6) 1.   │A   AP
2.   │ │B&~B   AP
3.   │ │B   Simp 2
4.   │ │~B   Simp 2
5.   │ │BvK   Add 3
6.   │ │K   DS 4,5
7.   │(B&~B)⊃K   CP 2-6
8. A⊃[(B&~B)⊃K] CP 1-7

(7) 1.   │(A⊃B)&~B   AP
2.   │A⊃B   Simp 1
3.   │~B   Simp 1
4.   │~A   MT 2,3
5. [(A⊃B)&~B]⊃~A   CP 1-4

```
(9) 1.          |(A⊃B)&(B⊃C)     AP
    2.          |A⊃B             Simp 1
    3.          |B⊃C             Simp 1
    4.          |A⊃C             HS 2,3
    9. [(A⊃B)&(B⊃C)]⊃(A⊃C)   CP 1-4


(10) 1.   |[(A⊃B)&(E⊃S)]&(AvE) AP
     2.   |(A⊃B)&(E⊃S)         Simp 1
     3.   |AvE                 Simp 1
     4.   |A⊃B                 Simp 2
     5.   |E⊃S                 Simp 2
     6.   |BvS                 CD 3, 4, 5
     7.{[(A⊃B)&(E⊃S)]&(AvE)}⊃(BvS) CP 1-6
```

39

Chapter 10

This chapter introduces a whole new type of rule, *replacement* rules. Make sure you understand the difference between these new rules and the inference rules of Chapters 7 and 8. Using replacement rules, many arguments can be proven valid that would not otherwise be provable. Thus, with the addition of replacement rules, our natural deduction system becomes an even more powerful logical tool.

Answers

Exercise 10.1

```
2.   5.  Comm  1
     6.  Assoc  5
     7.  Simp  2
     8.  DS  6,7
     9.  CD  3,4,8
    10.  DM  9

3.   5.  DM  1
     6.  DNeg  5
     7.  Simp  6
     8.  MP  2,7
     9.  DNeg  8
    10.  MT  3,9
    11.  MP  4,10
    12.  DM  11
```

Exercise 10.2 Part I.

```
(1)1.(AvB)⊃~E
   2. Bv(GvA)
   3. ~G   /~E
   4. Bv(AvG)      Comm 2
   5. (BvA)vG      Assoc 4
   6. Gv(BvA)      Comm 5
   7. BvA          DS 3,6
   8. AvB          Comm 7
   9. ~E           MP  1,8


(2)1.(H⊃S)v~R
   2. (AvB)vQ
   3. ~B&~S
   4. (AvQ)⊃R   /~H
   5. (BvA)vQ           Comm 2
   6. Bv(AvQ)           Assoc 5
   7. ~B                Simp 3
   8. AvQ               DS  6,7
```

```
9. R                      MP  4,8
10.~R v (H ⊃ S)           Comm 1
11.~~R                    DNeg 9
12. H ⊃ S                 DS 10,11
13. ~S                    Simp 3
14. ~H                    MT 13,12


(4)1.(AvB)⊃G
2. R&H
3. ~(~A&~B)/G
4. AvB           DM  3
5. G             MP  1,4


(8)1.A&(BvC)
2. (A&B)⊃H
3. (A&C)⊃O /H v O
4. (A&B)v(A&C) Dist  1
5. HvO           CD  2,3,4

(9)1.~(~Av~B)
2. B⊃H
3. H⊃S /S
4. ~~(~~A&~~B) DM 1
5. A&B           DNeg  4
6. B             Simp  5
7. H             MP  2,6
8. S             MP  3,7


(12)1.(G≡H)v~F
2.~(Av~F)
3.(G≡H)⊃P /P
4. ~~(~A &~~F)               DM 2
5. ~A & F                    DNeg  4
6. F                         Simp  5
7. ~F v (G ≡ H)              Comm  1
8. ~~F                       DNeg  6
9. G ≡ H                     DS  7, 8
10. P                        MP  3,9

(14)1.(P⊃Q)&(R⊃S)
2. (QvS)⊃~H
3. Pv(R&B)/~ H
4. P⊃Q                       Simp 1
5. R⊃S                       Simp  1
```

```
6.  (PvR)&(PvB)            Dist 3
7.  PvR                    Simp  6
8.  QvS                    CD   4,5,7
9.  ~H                     MP    2,8


(16) 1.Av(B&C)
2. ~B
3. ~AvE  /E
4. (AvB)&(AvC)             Dist 1
5. AvB                     Simp  4
6. BvA                     Comm  5
7. A                       DS   2,6
8. ~~A                     DNeg  7
9. E                       DS   3,8


(19)  1.~A
2. (A&B)v(H&S) / S
3. ~Av~B                   Add   1
4. ~(A&B)                  DM    3
5. H&S                     DS   2,4
6. S                       Simp  5


(21) 1.H⊃S
2. K⊃H
3. ~(A v S) /~(KvA)
4. ~A&~S                   DM   3
5. ~S                      Simp  4
6. ~H                      MT   1,5
7. ~K                      MT   2,6
8. ~A                      Simp  4
9. ~K&~A                   Conj  7,8
10.~(KvA)                  DM   9


(22) 1.~(AvB)
2. (~BvC)⊃(E&G)
3. S⊃~G   /~SvX
4. ~A&~B                   DM   1
5. ~B                      Simp  4
6. ~BvC                    Add  5
7. E&G                     MP   2,6
8. G                       Simp  7
9. ~~G                     DNeg  8
10. ~S                     MT   3,9
11. ~SvX                   Add   10


(24)1.QvR
2. [(PvQ)vR]⊃~S
3. H⊃S /~H
4. (QvR)vP        Add  1
```

```
5.  Pv(QvR)          Comm   4
6.  (PvQ)vR          Assoc  5
7.  ~S               MP   2,6
8.  ~H               MT   3,7


(26)1. (AvB)⊃~(HvE)
2.  Av(B&H)
3.  (A&~E)⊃(P&Q)  /Q v X
4.  (AvB)&(AvH)                  Dist  2
5.  AvB                          Simp  4
6.  ~(HvE)                       MP  1,5
7.  ~H&~E                        DM   6
8.  AvH                          Simp  4
9.  HvA                          Comm 8
10. ~H                           Simp  7
11. A                            DS  9,10
12. ~E                           Simp  7
13. A&~E                         Conj 11,12
14. P&Q                          MP  3,13
15. Q                            Simp 14
16. QvX                          Add  15


(28)1 RvS
2.  ~(A&R)
3.  ~(A&S)   / ~A
4.  ~Av~R                        DM   2
5.  ~Av~S                        DM   3
6.  (~Av~R)&(~Av~S)              Conj 4,5
7.  ~Av(~R&~S)                   Dist 6
8.  ~(~R&~S)                     DM   1
9.  (~R&~S) v ~A                 Comm 7
10. ~A                           DS  8,9


(29)1. (AvB)⊃~E
2.  (H&A)v(B&E)   /A
3. [(H&A)vB]&[(H&A)vE]           Dist 2
4.  (H&A)vB                      Simp  3
5.  Bv(H&A)                      Comm  4
6.  (BvH)&(BvA)                  Dist  5
7.  BvA                          Simp  6
8.  AvB                          Comm  7
9.  ~E                           MP  1,8
10. (H&A)vE                      Simp  3
11. Ev(H&A)                      Comm 10
12. H&A                          DS  9,11
13. A                            Simp 12
```

Exercise 10.2 Part II.

```
(2)1.~(AvB)
2.Bv~E
3.J⊃E / ~J
4.  ~A&~B        DM 1
5.  ~B           Simp 4
6.  ~E           DS 2,5
7.  ~J           MT 3,6


(3)1. Av(BvD)
2.  ~(DvC)
3.  ~(AvB)vS /S
4.  ~D&~C                    DM 2
5.  (AvB)vD                  Assoc 1
6.  Dv(AvB)                  Comm 5
7.  ~D                       Simp 4
8.  AvB                      DS 6,7
9.  ~~(AvB)                  DNeg 8
10. S                       DS 3,9


(5) 1.A&(BvS)
2.  J⊃~(A&S)
3.  ~J⊃~K
4.  K v G
5.  ~(A&B)   / G
6.  (A&B)v(A&S)             Dist 1
7.  A&S                     DS 6, 5
8.  ~~(A&S)                 D Neg 7
9.  ~J                      MT 2, 8
10. ~K                      MP 3, 9
11. G                       DS 4, 10
```

Exercise 10.3

```
2.  5.   Equiv  1
    6.   Simp  5
    7.   MP  2,6
    8.   MP  4,7
    9.   Equiv  1
   10.   DM   3
   11. DNeg 10
   12.   DS  9,11
   13.   Conj  8,12

3.  5.   DM   3
    6.  DNeg 5
    7.  DNeg 6
```

44

```
8.   CD   1,2,7
9.   Taut  8
10.  MP   4,9
```

Exercise 10.4 Part I

```
(2) 1.~A⊃B
2.  A⊃E
3.  B⊃S
4.  (EvS)⊃X     / X
5.  ~~AvB        Imp 1
6.  AvB          DNeg 5
7.  EvS          CD  2,3,6
8.  X            MP  4,7


(3) 1.  J⊃I
2.  (~I⊃~J)⊃S
3.  F⊃~S / ~F
4.  ~I⊃~J        Trans  1
5.  S            MP  2,4
6.  ~~S          DNeg 5
7.  ~F           MT   3,6


(6)1.  A⊃B
2.  A⊃~B
3.  ~A⊃G     / G
4.  ~~B⊃~A       Trans  2
5.  B⊃~A         DNeg 4
6.  A⊃~A         HS   1,5
7.  ~Av~A        Imp  6
8.  ~A           Taut  7
9.  G            MP   3,8


(8)1.A⊃(B⊃C)   /B⊃(A⊃C)
2.(A&B)⊃C       Exp  1
3.(B&A)⊃C       Comm  2
4.B⊃(A⊃C)       Exp  3


(9)1.  H⊃(A⊃E)
2.  (Ev~A)⊃S  /H⊃S
3.  (~AvE)⊃S     Comm  2
4.(A⊃E)⊃S        Imp  3
5.  H⊃S          HS   1,4
```

(12) 1. B≡G
2. (B&G)⊃Z
3. (~B&~G)⊃Z /Z
4. (B&G)v(~B&~G)   Equiv 1
5. ZvZ   CD 2,3,4
6. Z   Taut 5

(14) 1. A⊃(B&C)
2. B⊃(C⊃S) /A ⊃ S
3. (B&C)⊃S   Exp 2
4.   A⊃S   HS   1,3

(16) 1. ~[(J⊃I)&(I⊃J)]
2. (A&S)⊃(J≡I)
3. A   /~S
4. ~(J≡I)   Equiv 1
5. ~(A&S)   MT   2,4
6. ~Av~S   DM   5
7. ~~A   DNeg 3
8. ~S   DS   6,7

(17) 1. (A⊃B)⊃[(TvI)&(H≡J)]
2. (TvI)⊃[(H≡J)⊃X] /(A⊃B)⊃X
3. [(TvI)&(H≡J)]⊃X   Exp 2
4. (A⊃B)⊃X   HS   1,3

(18) 1. A⊃(B&E) /(A⊃B)&(A⊃E)
2. ~Av(B&E)   Imp 1
3. (~AvB)&(~AvE)   Dist 2
4. (A⊃B)&(A⊃E)   Imp 3 (twice)

(19) 1. H⊃~H
2. ~(H&S)⊃G   /~(Hv~G)
3. ~Hv~H   Imp 1
4. ~H   Taut 3
5. ~Hv~S   Add 4
6. ~(H&S)   DM   5
7. G   MP 2,6
8. ~H&G   Conj. 4,7
9. ~(~~H v~G)   DM 8
10. ~(Hv~G)   DNeg 9

(22) 1. ~A⊃S
2. ~E⊃S
3. ~(A&E) /SvH

```
4.  ~Av~E                          DM   3
5.  SvS                            CD   1,2,4
6.  S                              Taut 5
7.  SvH                            Add  6


(24) 1.  Jv(I&S)
2.  J⊃S    / S
3.  (JvI)&(JvS)                     Dist  1
4.  JvS                             Simp  3
5.  ~~J v S                         DNeg  4
6.  ~J⊃S                            Imp   5
7.  ~S⊃~~J                          Trans 6
8.  ~S⊃J                            DNeg  7
9.  ~S⊃S                            HS    2,8
10. ~~SvS                           Imp   9
11. SvS                             DNeg  10
12. S                              Taut  11


(26) 1.A⊃B  /A⊃(BvE)
2.  ~AvB                            Imp   1
3.  (~AvB)vE                        Add   2
4.  ~Av(BvE)                        Assoc 3
5.  A⊃(BvE)                         Imp   4


(27) 1.A⊃~(B⊃G)
2.  AvG   /A≡~G
3.  ~Av~(B⊃G)                       Imp   1
4.  ~Av ~(~BvG)                     Imp   3
5.  ~Av ~~(~~B&~G)                  DM    4
6.  ~Av(B&~G)                       DNeg  5
7.  (~AvB)&(~Av~G)                  Dist  6
8.  ~Av~G                           Simp  7
9.  A⊃~G                            Imp   8
10. GvA                             Comm  2
11. ~~GvA                           DNeg  10
12. ~G⊃A                            Imp   11
13.(A⊃~G)&(~G⊃A)                    Conj  9,12
14. A ≡ ~G                          Equiv 13


(31)1. A≡B
2.  AvB
3.  A⊃(B⊃E) / E
4.(A⊃B)&(B⊃A)                       Equiv 1
5.  ~~AvB                           DNeg  2
6.  ~A⊃B                            Imp   5
7.  B⊃A                             Simp  4
```

```
8.  ~A⊃A                        HS   6,7
9.  ~~AvA                       Imp  8
10. AvA                         DNeg 9
11. A                           Taut  10
12. B⊃E                         MP   11,3
13. ~B⊃~~A                      Trans 6
14. ~B⊃A                        DNeg 13
15. A⊃B                         Simp  4
16. ~B⊃B                        HS   14,15
17. ~~BvB                       Imp  16
18. BvB                         DNeg 17
19. B                           Taut  18
20. E                           MP   12,19


(32) 1. A⊃(B&H)
2. (HvS)⊃Z  /A⊃Z
3. ~Av(B&H)                     Imp 1
4. (~AvB)&(~AvH)                Dist 3
5. ~AvH                         Simp 4
6. (~AvH)vS                     Add 5
7. ~Av(HvS)                     Assoc 6
8. A⊃(HvS)                      Imp 7
9. A⊃Z                          HS 2, 8
```

Exercise 10. 4 Part II

```
(1) 1. ~A⊃~B
2. (~BvA)⊃S  / S
3. B⊃A              Trans 1
4. ~BvA             Imp 3
5. S                MP 2, 4


(2) 1. (A&B)⊃E
2. A  / B⊃E
3. A⊃(B⊃E)          Exp 1
4. B⊃E              MP 2, 3


(3)  1. A≡E
2. E≡B  / ~B⊃~A
3. (A⊃E)&(E⊃A)      Equiv 1
4. (E⊃B)&(B⊃E)      Equiv 2
5. A⊃E              Simp 3
6. E⊃B              Simp 4
7. A⊃B              HS 5, 6
8. ~B⊃~A            Trans 7
```

Exercise 10.5.

(1) Pat goes home three hours early:

```
1.  ~(A&B)
2.  (~Bv~C) &~(~B&~C)
3. ~CvD
4.  ~A⊃~E
5.  ~(~Ev~K)
6. D⊃P  / P
7.  E&K           DM 5
8. E              Simp 7
9. ~~E            DNeg 8
10. ~~A           MT 4,9
11. ~Av~B         DM 1
12. ~B            DS 10,11
13. ~(~B&~C)      Simp 2
14. BvC           DM 13
15. C             DS 12, 14
16. ~~C           DNeg 15
17. D             DS 3,16
18. P             MP 6,17
```

2.(a) To derive a formula ~P from the corresponding formulas P⊃Q and ~Q, transpose P⊃Q and derive ~Q⊃~P, then apply MP to this and ~Q to prove ~P.

(c) To derive a formula P⊃R from the corresponding formulas P⊃Q and Q⊃R, use the Indirect Proof method: Assume ~(P⊃R). Apply Imp to the assumed premise to derive ~(~PvR). Next, apply DM and DNeg to this to prove P&~R. Then apply Simp to this to derive P and apply Simp again to derive ~R. Then apply MP to P and P⊃Q to prove Q; apply MT to ~R and Q⊃R to prove ~Q, conjoin Q and ~Q for your contradiction, discharge the assumption, and assert P⊃R.

Chapter 11

This chapter combines the rules and techniques of Chapters 7, 8, 9, and 10. If you thought those chapters were fun, you'll think this one's a real blast.

Answers

Exercise 11.1

```
(1) 1. J⊃(A&B)
2.  (AvE)⊃R
3.  EvJ       / R
4.  |  ~R           AP
5.  |  ~(AvE)       MT  2,4
6.  |  ~A&~E        DM  5
7.  |  ~E           Simp  6
8.  |  J            DS  3,7
9.  |  A&B          MP  1,8
10. |  A            Simp  9
11. |  ~A           Simp  6
12. |  A & ~A       Conj 10,11
13. R              IP  4-12
```

```
(2) 1. (A⊃A)⊃B
2. (BvG)⊃C  / C
3.  |  ~C           AP
4.  |  ~(BvG)       MT  2,3
5.  |  ~B&~G        DM  4
6.  |  ~B           Simp  5
7.  |  ~(A⊃A)       MT  6,1
8.  |  ~(~AvA)      Imp  7
9.  |  ~~(~~A&~A)   DM  8
10. |  A&~A         DNeg 9
11. C              IP  3-9
```

```
(5) 1. A⊃B
2. I⊃~B
3. ~K⊃(A&I)  /K v S
4.            |  ~(KvS)        AP
5.            |  ~K&~S         DM  4
6.            |  ~K            Simp  5
7.            |  A&I           MP  3,6
8.            |  A             Simp  7
9.            |  I             Simp  7
10.           |  B             MP  1,8
11.           |  ~B            MP  2,9
```

```
12.              │ B & ~B           Conj 10,11
13.  KvS                           IP  4-12


(6)1. J⊃(I⊃R)
2.   J⊃I
3.   ~Kv(RvJ)  /~K v R
4.              │ ~(~KvR)          AP
5.              │ ~~(~~K&~R)       DM 4
6.              │ ~~K &~R          DNeg  5
7.              │ ~~K              Simp 6
8.              │ RvJ              DS  3,7
9.              │ ~R               Simp  6
10.             │ J                DS  8,9
11.             │ I⊃R              MP  1,10
12.             │ ~I               MT  9,11
13.             │ I                MP  2,10
14.             │ I & ~I           Conj 12,13
15. ~KvR                          IP  4-14


(9)1. (JvI)⊃(E&H)
2. (HvA)⊃(Bv~E)
3. (BvK)⊃~(J&H)  /~J
4.              │ ~~J              AP
5.              │ J                DNeg 4
6.              │ JvI              Add   5
7.              │ E&H              MP  1,6
8.              │ H                Simp  6
9.              │ ~~H              DNeg 8
10.             │ HvA              Add   8
11.             │ Bv~E             MP  2,10
12.             │ E                Simp  7
13.             │ ~~E              DNeg 12
14.             │ ~E v B           Comm 11
15.             │ B                DS  13,14
16.             │ BvK              Add  15
17.             │ ~(J&H)           MP  3,16
18.             │ ~Jv~H            DM  17
19.             │ ~H v ~J          Comm 18
20.             │ ~J               DS  9,19
21.             │ J & ~J           Conj 5, 20
22.  ~J                            IP  4-21



(11)1. A⊃(S⊃G)
2. A⊃S
3. I⊃(AvG)  /Gv~I
4.              │ ~(Gv~I)          AP
5.              │ ~G&~~I           DM  4
```

```
6.              | ~G & I              DNeg 5
7.              | I                   Simp 6
8.              | AvG                 MP  3,7
9.              | ~G                  Simp  6
10.             | GvA                 Comm 8
11.             | A                   DS   9,10
12.             | S                   MP  11,2
13.             | S⊃G                 MP  1,11
14.             | G                   MP  12,13
15.             | G & ~G              Conj 9,14
16.   Gv~I                            IP   4-15


(13) 1. ~[(A⊃B)v(B⊃A)]
2. Ev[Sv(B⊃A)]
3. ~H⊃~(SvE) / H
4.              | ~H                  AP
5.              | ~(S v E)            MP 3,4
6.              | ~S & ~E             DM 5
7.              | ~E                  Simp 6
8.              | S v (B ⊃A)          DS 2, 7
9.              | ~S                  Simp 6
10             | B ⊃ A               DS 8, 9
11.            | ~(A ⊃B) & ~(B ⊃ A)  DM 1
12.            | ~(B ⊃ A)            Simp 11
13.            | (B ⊃A) & ~(B ⊃ A)   Conj 10, 12
14. H                                IP 4-13


(15) 1. A⊃(B&C)
2. (BvC)⊃W
3. ~C ⊃(A&W) / W
4.    | ~W     AP
5.    | ~(B v C)          MT 2, 4
6.    | ~B & ~C           DM 5
7.    | ~C                Simp 6
8.    | A & W             MP 3, 7
9.    | W                 Simp 8
10    | W & ~W            Conj 9, 4
11.   W                   IP 4-10


(16) 1. A⊃B
2. C⊃D
3. (BvD)⊃H
4. ~H /~(AvC)
5.        | ~~(AvC)       AP
6.        | AvC           DNeg 5
7.        | BvD           CD 1, 2, 6
```

52

```
8.        |H               MP 3, 7
9.        |H&~H            Conj 4, 8
10. ~(A v C)               IP 5-9


(18) 1.~[(A&~B)&~C]
2.~(BvC)   /~A
3.    |~~A                 AP
4.    |A                   D Neg 3
5.    |~(A & ~B) v ~~C     DM 1
6.    |(~A v ~~B) v ~~C    DM 5
7.    |(~A v B) v C        D Neg 6
8.    |~B & ~C             DM 2
9.    |~C                  Simp 8
10    |C v (~A v B)        Comm 7
11    |~A v B              DS 9, 10
12    |A ⊃ B               Imp 11
13    |B                   MP 4, 12
14    |~B                  Simp 8
15    |B & ~B              Conj 13, 14
16    ~A                   IP 3-15


(20) 1. A&(H&B)
2. [(JvH)&X]⊃E
3. (C⊃~E)
4. X / ~C
5.    |~~C                 AP
6.    |C                   DNeg 5
7.    |~E                  MP 3, 6
8.    |~[(J v H) & X]      MT 2, 7
9.    |~(J v H) v ~X       DM 8
10    |(~J & ~H) v ~X      DM 9
11    |~X v (~J & ~H)      Comm 10
12    |~~X                 DNeg 4
13    |~J & ~H             DS 11, 12
14    |~H                  Simp 13
15    |H & B               Simp 1
16    |H                   Simp 15
17    |H & ~H              Conj 14, 16
18.   ~C                   IP  5-17


(21)1.~AvB
2. B⊃C
3. H⊃(A&~C)
4. HvD / D
5.    |~D                  AP
6.    |D v H               Comm 4
7.    |H                   DS 5, 6
```

```
8.   | A & ~C           MP 3, 7
9.   | A                Simp 8
10   | ~~A              DNeg 9
11   | B                DS 1, 10
12   | C                MP 2, 11
13   | ~C               Simp 8
14   | C & ~C           Conj 12, 13
15.  D                  IP 5-14
```

Exercise 11.3

```
(2)1. A⊃(B&C)
2.  B⊃(A&G)      / A≡B
3.        | A           AP
4.        | B&C         MP  1,3
5.        | B           Simp  4
6.  A⊃B                 CP   3-5
7.        | B           AP
8.        | A&G         MP  2,7
9.        | A           Simp  8
10.  B⊃A                CP   7-9
11.  (A⊃B)&(B⊃A)  Conj  6,10
12.  A ≡ B              Equiv  11
```

```
(5)1.(A&B)⊃C
2.A⊃B / A⊃C
3.      | A            AP
4.      | B            MP  2,3
5.      | A&B          Conj  3,4
6.      | C            MP  1,5
7.  A⊃C                CP  3-6
```

```
(6)1.I⊃Z
2. Z⊃A /I⊃(Z & A)
3.     | I             AP
4.     | Z             MP 1,3
5.     | A             MP 2,4
6.     | Z & A         Conj 4,5
7. I ⊃ (Z & A)         CP 3-6
```

```
(7)1.(JvI)⊃S
2.~ J⊃~A /A⊃(~S⊃T)
3.              | A            AP
4.              | ~~A          DNeg 3
5.              | ~~J          MT 2,4
6.              | J            DNeg 5
```

54
```
```

7.		JvI	Add 6
8.		S	MP 1,7
9.		SvT	Add 8
10.		~~S v T	DNeg 9
11.		~S⊃T	Imp 10
12.	A⊃(~S⊃T)		CP 3-11

(8) 1. H⊃(G⊃J)
2. G⊃(J⊃W) /H⊃(~W⊃~G)

3.		H	AP	
4.			~W	AP
5.			G ⊃ J	MP 1, 3
6.			(G & J) ⊃ W	Exp 2
7.			~(G & J)	MT 4, 6
8.			~ G v ~J	DM 7
9.			~J v ~G	Comm 8
10			J ⊃ ~G	Imp 9
11			G ⊃ ~G	HS 5, 10
12			~G v ~G	Imp 11
13			~G	Taut 12
14		~W⊃~G		CP 4-13
15.	H⊃(~W⊃~G)		CP 3-14

(9) 1. A⊃[(JvI)⊃(M&N)]
2. (NvS)⊃G / A⊃(I⊃G)

3.		A AP		
4.			I	AP
5.			(JvI)⊃(M&N)	MP 1,3
6.			IvJ	Add 4
7.			JvI	Comm 6
8.			M&N	MP 7,5
9.			N	Simp 8
10.			NvS	Add 9
11			G	MP 2,10
12.		I⊃G		CP 4-11
13.	A⊃(I⊃G)			CP 3-12

Exercise 11.5

(1) 1.

1.		~~[(A⊃~A)&(~A⊃A)]	AP
2.		[(A⊃~A)&(~A⊃A)]	DNeg 1
3.		A⊃~A	Simp 2
4.		~Av~A	Imp 3
5.		~A	Taut 4
6.		~A⊃A	Simp 2

55

```
7.              | ~~AvA                    Imp  6
8.              | AvA                      DNeg 7
9.              | A                        Taut 8
10.             | A&~A                     Conj 9,5
11. ~[(A⊃~A)&(~A⊃A)]                       IP  1-10

(2)1.           | ~(B&G)                   AP
2.              | ~Bv~G                    DM  1
3.   ~(B&G)⊃(~Bv~G)                        CP  1-2

(5)1.           | A≡B                      AP
 2.             | (A⊃B)&(B⊃A)              Equiv 1
 3.  (A≡B)⊃[(A⊃B)&(B⊃A)]                   CP 1-2

(6)1.           | (J&I)⊃I                  AP
2.              | J⊃(I⊃I)                  Exp  1
3.  [(J&I)⊃I]⊃[J⊃(I⊃I)]                    CP 1-2

(7)1.           | ~[~(A&~A)v(B&~B)]        AP
2.              | ~~(A&~A)&~(B&~B)         DM  1
3.              | (A&~A)&~(B&~B)           DNeg 2
4.              | A&~A                     Simp 3
5.~(A&~A)v(B&~B)                           IP  1-4

(9)1.           | ~[~J⊃(J⊃I)]              AP
2.              | ~[~~Jv(J⊃I)]             Imp 1
3.              | ~[Jv(J⊃I)]               DNeg 2
4.              | ~J&~(J⊃I)                DM 3
5.              | ~J                       Simp  4
6.              | ~(J⊃I)                   Simp  4
7.              | ~(~JvI)                  Imp  6
8.              | ~~J&~I                   DM  7
9.              | J&~I                     DNeg 8
10.             | J                        Simp  9
11.             | J & ~J                   Conj 5,10
12.~J⊃(J⊃I)                                IP  1-11

(12) 1.         | ~[(J⊃I)v(I⊃J)]           AP
2.              | ~(J⊃I)&~(I⊃J)            DM  1
3.              | ~(J⊃I)                   Simp  2
4.              | ~(~JvI)                  Imp  3
5.              | ~~J&~I                   DM  4
6.              | J&~I                     DNeg 5
7.              | ~(I⊃J)                   Simp  2
8.              | ~(~IvJ)                  Imp  7
9.              | ~~I&~J                   DM  8
```

```
10.              I&~J              DNeg 9
11.              J                 Simp 6
12.              ~J                Simp 10
13.              J & ~J            Conj 11,12
14.  (J⊃I)v(I⊃J)                   IP  1-13


(13)1.           ~[(J⊃~J)v(~J⊃J)]  AP
2.               ~(J⊃~J)&~(~J⊃J)   DM 1
3.               ~(J⊃~J)           Simp 2
4.               ~(~Jv~J)          Imp  3
5.               ~~J               Taut  4
6.               J                 DNeg 5
7.               ~(~J⊃J)           Simp 2
8.               ~(~~JvJ)          Imp  7
9.               ~(JvJ)            DNeg 8
10.              ~J                Taut  9
11.              J & ~J            Conj 6,10
12.    (J⊃~J)v(~J⊃J)               IP  1-11

(15)1.           ~(A&~B)&~B        AP
2.               ~(A&~B)           Simp  1
3.               ~Av~~B            DM  2
4.               ~~Bv~A            Comm 3
5.               Bv~A              DNeg 4
6.               ~B                Simp  1
7.               ~A                DS  5,6
8.  [~(A&~B)&~B]⊃~A                CP  1-7


(17)1.           ~[(~A⊃B)v(A⊃E)]   AP
2.               ~(~A⊃B)&~(A⊃E)    DM  1
3.               ~(~~AvB)&~(~AvE)  Imp  2
4.               ~(AvB)&~(~AvE)    DNeg 3
5.               (~A&~B)&(~~A&~E)  DM  4 (twice)
6.               (~A&~B)&(A&~E)    DNeg 5
7.               ~A&~B             Simp  6
8.               ~A                Simp  7
9.               A&~E              Simp  6
10.              A                 Simp  9
11.              A & ~A            Conj 8,10
12.  (~A⊃B)v(A⊃E)                  IP  1-11

(18)1. A  AP
2.               S                 AP
3.               SvS               Add 2
4.               S                 Taut  3
5.      S⊃S                        CP  2-4
6.      ~SvS                       Imp  5
```

```
 7.        | Sv~S                                      Comm   6
 8.        | A&(Sv~S)                                  Conj   1,7
 9. A⊃[A&(Sv~S)]                                       CP     1-8
10.                    | A&(Sv~S)                      AP
11.                    | A                             Simp   10
12.    [A&(Sv~S)]⊃A                                    CP     10-11
13.    {A⊃[A&(Sv~S)]}&{[A&(Sv~S)]⊃A}  Conj 9, 12
14.    A≡[A&(Sv~S)]                                    Equiv  13


(19)1.        | ~[(A⊃B)v(~B⊃A)]                        AP
 2.           | ~(A⊃B)&~(~B⊃A)                         DM 1
 3.           | ~(~AvB)&~(~~BvA)                       Imp 2
 4.           | ~(~AvB)&~(BvA)                         DNeg 3
 5.           | ~(~AvB)                                Simp 4
 6.           | ~~A&~B                                 DM  5
 7.           | A&~B                                   DNeg 6
 8.           | ~(BvA)                                 Simp 4
 9.           | ~B&~A                                  DM  8
10.           | ~A                                     Simp 9
11.           | A                                      Simp 7
12.           | A&~A                                   Conj 10,11
13.    (A⊃B)v(~B⊃A)                                    IP   1-12


(20)1.        | A⊃B  AP
 2.           |                  | A⊃~B        AP
 3.           |                  | ~~B⊃~A      Trans  2
 4.           |                  | B⊃~A        DNeg 3
 5.           |                  | A⊃~A        HS 1,4
 6.           |                  | ~Av~A       Imp 5
 7.           |                  | ~A          Taut  6
 8.           | (A⊃~B)⊃~A                      CP   2-7
 8.    (A⊃B)⊃[(A⊃~B)⊃~A]                        CP   1-8


(22) 1.       | ~(A&B) & B                             AP
 2.           | ~(A&B)                                 Simp 1
 3.           | ~Av~B                                  DM 2
 4.           | B                                      Simp 1
 5.           | ~Bv~A                                  Comm 3
 6.           | ~~B                                    DNeg 4
 7.           | ~A                                     DS 5,6
 8.[~(A&B)&B]⊃~A                                       CP 1-7
```

Chapter 12

Human disagreement is sometimes due to a failure of communication. Sometimes this happens when one person means one thing by a word and the other person means something else entirely. The two individuals are "talking past each other." In such situations, it helps if the parties to the disagreement *define* their terms and explain exactly what they mean. The art of definition--the ability to adequately define one's terms--is an important skill.

Answers

Exercise 12.1

1. A lexical definition might do. A theoretical definition would be better.

3. Lexical definition.

6. While on patrol behind German lines, Sergeant Saunders (of "Combat" fame) and his men capture a group of German soldiers and tell them, "Our medic is now going to take your arms from you." The soldiers act frightened. In this short plot that never made it into the 1960's series "Combat," *arms* is used ambiguously since it is not clear whether the medic is going to amputate limbs or confiscate guns.

c. A rich person is anyone who earns over $200, 000 per year or who has a net worth of over one million dollars. Inadequate, for someone whose income is $190, 000 and whose net worth is $900,000 is certainly rich.
e. A liberal is a person who believes that the government should not confine itself only to protecting life, liberty, and property, but should also provide welfare for the poor, regulate the economy, support the arts, and use the tax system to make incomes more equal.
g. Cold weather is any weather below 35 degrees. Inadequate. In may cases, 40 degree weather is cold.
h. Poverty is a condition in which one does not have either (a) adequate food to eat, or (b) adequate clothes to wear, or (c) adequate housing. Inadequate. There are people who live in poverty but who get government assistance and consequently have plenty to eat, plenty of clothes, and adequate housing in a decent building. Yet they are poor.

9. a. Socialism. A condition in which people live in harmony because the state owns and controls the major means of production and consequently ends the exploitation of the many by the few and also ensures a more or less equal distribution of wealth.
b. Capitalism. A condition in which people are free to chart their own course in life.
c. Religion. Union with God, the source of all goodness and being.
d. Communism. A state of affairs that will be reached when all exploitation has been abolished and each person receives equal concern and respect as a human being, regardless of productive contribution to society.
e. Intellectual. Someone who has attained a superior knowledge of some aspect of the world.
h. An altruist helps other people and consequently makes the world a better place.
i. Selfishness is an irrational attachment to one's own needs and desires, to the detriment of everybody else.

b. A cleaning agent made from chemical compounds rather than natural fats or oils.

e. A visible body of small droplets of water.

f. A mixture of smoke and fog.

g. The smallest particle of an element having all the properties of the element.

11. b. lexical

d. stipulative, precising

f. lexical

g. persuasive

i. theoretical

k. lexical

l. lexical

n. lexical

p. persuasive

q. persuasive

Exercise 12.2

2. a. musical instruments, orchestral instruments, woodwinds, saxophones, alto saxophones, old alto saxophones.

c. sliced American cheese, American cheese, cheese, milk derivatives, dairy products.

4. a. Astronaut. Intension: A person trained to operate spacecraft.

Part of extension: Neil Armstrong, Buzz Aldrin, John Glenn.

b. Famous. Intension: Possessing a widespread reputation.

Part of extension: Bill Clinton, Jerry Seinfeld, Bob Hope.

Part of extension(for humans): Vegetables, fruits, cheese. (Hamburgers? Hot dogs?)

e. Pope. Intension: Bishop of Rome, head of the Roman Catholic Church.

Part of extension: Leo XIII (Pope from 1878-1903), John XXIII, John Paul.

Part of extension: Billy Graham, Jesse Jackson, Martin Luther King, Jerry Falwell

g. Politician. Intension: One who holds or seeks political office.

Part of extension: Bill Clinton, Ronald Reagan, George Bush, Al Gore, Richard Nixon.

5. b. Communist: Supporter of communism. Influenced by the philosophy of Karl Marx.

c. Snake: Elongated, limbless, scaly reptile. (Or: treacherous, untrustworthy person.)

e. Conservative: Person disposed toward the preservation of existing institutions, distrustful of revolutions.

g. Science: The body of knowledge gained by the employment of the scientific method.

Exercise 12.3

2. E	8.E
3. I	10.I
6. I	

Exercise 12.4

1. enumerative	8. synonymous	
4. operational	9. enumerative	
5. operational	13. analytical	14. enumerative

Exercise 12.5

1.a. uninteresting
b. irritating
c. humorous

2. a. A musical group such as the Beatles, the Doors, or the Moody Blues.
b. A person such as Senator Patty Murray, Senator Slade Gorton.

3. b. To show esteem, to honor, to value, to defer to.
e. Instructions, usually recorded in a magnetic medium, that make a computer function in a particular way.

4. a. If you wonder whether person x is funny, gather ten persons at random and have person x tell them a joke. If at least eight laugh, then person x is funny.
d. If x floats when thrown into a lake, then x is buoyant.
e. If x causes the needle of a current detector to move, then x is electrically charged.

Exercise 12.6

1. genus: cats
difference: male

4. genus: men
difference: unmarried

5. genus: mammals
difference: insect-eating

8. genus: integers
difference: integers whose only divisors are themselves and 1.

Exercise 12.7.

a. Rule 2: too broad

d. Rule 3: too negative.

e. Rule 3: too negative.

h. Rule 3: too negative.

i. Rule 5. plays on the emotions.

k. Rule 5. plays on the emotions.

n. Rule 6. circular

o. Rule 6. circular   p. Rule 2: too broad

Chapter 13

A *fallacy* is an error in reasoning that may nevertheless look like a correct argument. Nowadays, we are bombarded with arguments from every direction--arguments urging us to buy this product or that one, arguments urging us to vote for this or that political candidate, and so on. And not all of these arguments are good ones. A critical thinker must therefore always be on the lookout for logical fallacies. A very large number of different fallacies have been catalogued by logicians. This chapter introduces many of the most important fallacies.

Answers

The following answers are my interpretations. However, many of these fallacies may be interpreted differently by different people, and you may disagree with a few of my suggested answers.

Exercise 13.2

2. Ad hominem
3. circumstantial ad hominem
6. Ad populum
7. Ad misericordiam
8. begging the question
11. Circumstantial ad hominem
12. Ad populum

15. Begging the question
16. Begging the question
18. Tu quoque
20. Tu quoque
22. Circumstantial ad hominem
24. Appeal to ignorance

Exercise 13.4

2. accident
4. weak analogy
6. false dilemma
8. post hoc ergo propter hoc
9. accident
12. false dilemma
14. false dilemma

17. false dilemma
19. accident
21. accident
23. hasty generalization
25. accidental correlation
27. accident

Exercise 13. 6

3. division
4. division
5. division
7. division
9. division
10. division
12. composition
13. equivocation
14. amphiboly

16. division
19. equivocation
20. composition
22. division
23. composition
25. division
26. composition
29. division
30. division

Chapter 14

We often reason about *groups*. For instance, a biologist might begin a lecture by saying, "All living things contain DNA." A doctor might argue for the proposition that no human beings are naturally immune to AIDs. Or, in a moment of logical frustration, you might mutter, "Some logic problems cannot be solved, " and then you might conclude something from this. In this chapter, you will study some of the logic pertaining to this type of reasoning--reasoning about groups of things.

In this chapter, you will translate categorical sentences about groups into certain standardized formats. Translating these sentences into standardized forms fosters a deeper understanding of English grammar, it helps one write more precisely, and it also clarifies the logic of statements about groups.

This unit (Chapters 14 and 15) introduces a historically important branch of logic. Over 2,000 years ago, students in Aristotle's university studied this logical theory, as did students in Europe's first universities during the Middle Ages. This unit therefore introduces you to an important part of the intellectual history of Western Civilization. Graduate school entrance exams, civil service exams, and cognitive skills exams for jobs often test logical abilities using statements and problems drawn from the classic categorical logic of this unit.

Answers

Exercise 14.1

2. E   Universal negative
3. I   Particular affirmative

Exercise 14.2

2.	a.U		6.a.	F
	b.U		b.	T
	c.T		c.	T

4.	a.U		7.a.	F
	b.T		b.	U
	c.U		c.	U

Exercise 14.3

b.	Valid	m. Invalid
d.	Valid	n. Invalid
f.	Valid	p. Invalid
I.	Invalid	

Exercise 14.4

Part I	Part II
2. subcontraries	3. No extraterrestrials are green.
4. contradictories	4. Some extraterrestrials are not green.

Exercise 14.5

I. b. All particles of iron are particles that accept a magnetic charge.
    c. All teenagers are persons who are naturally full of energy.
    f. All truly religious people are charitable persons.
    g. All velvet Elvis paintings are beautiful things.
    h. All times Joe drives his car are times that the car breaks down.
    l. All persons admitted to the party are paid guests.
    m. All times involving weather are times it rains.

2. b. No particles of iron are particles that accept a magnetic charge.
    c. No teenagers are persons who are naturally full of energy.
    d. No heroin users are self-destructive persons.
    g. No times that Joe drives the car are times the car breaks down.
    h. No hamburgers are good sources of cholesterol.

    l. No humans are creatures who are selfish.
    m. No students are rich persons.

3. b. Some socialists are egalitarians.
    c. Some Democrats are persons from the South.
    f. Some Communists are Scrabble players.
    g. Some skateboarders are old persons.

4. b. Some teenagers are not energetic persons.
    c. Some cats are not mean animals.
    e. Some homes are not places that have VCR's.
    h. Some times when weather occurs are not times it rains.
    i. Some places where people have lived are not places that are sacred.

Exercise 14.6

1. b. No creepy things are spiders. (Equivalent.)

2. b. All Corvairs are non-safe vehicles. (Equivalent)
    d. Some Corvairs are non-safe vehicles. (Equivalent)

3. b. No non-honest persons are non-politicians.
    c. Some non-honest persons are non-politicians.

Chapter 15

The central focus of this chapter is the Venn diagram method. In this chapter, you will learn to transfer information from categorical statements onto a diagram called a "Venn diagram." By examining the diagrams, you will be able to literally *see* whether or not a categorical syllogism is valid. After you learn this method, you might take a moment to contrast the visual nature of the Venn diagrams with the more abstract nature of the natural deduction method.

This chapter also introduces:

1. Sorites
2. Refutation by logical analogy
3. Enthememes

In the Venn diagrams of this chapter, the middle term is always assigned the bottom circle, the minor term gets the top left circle, and the major term always gets the top right circle. Although the text does not explicitly require you to follow this order when constructing your figures, you may wish to explicitly make this the standard order so that all your Venn diagrams look alike.

Also, when constructing your diagrams, remember to fill in all shading before placing any X's.

Answers

Exercise 15.1

1. b.   III-4
   c.   EIO-4
   f.   IAA-4
   h.   OAO-4

1.

c.	major term:	cold blooded creatures
	minor term:	mammals
	middle term:	fish

g.	major term:	things that are nutritious
	minor term:	Pink's hot dogs
	middle term:	things that are delicious

3. c. invalid
     d. invalid
     g. valid
     h. invalid

5. No guitarists are poets.
     All accountants are guitarists.
     No accountants are poets.

7. All cats are adorable critters.
   No cats are menacing maniacs.
   No menacing maniacs are adorable critters.

8. Some vagabonds are happy hobos.
   All happy hobos are ex-physicists.
   Some ex-physicists are vagabonds.

10. All baristas are ambidextrous persons.
   Some baristas are not cooks.
   Some cooks are not ambidextrous persons.

Exercise 15.2

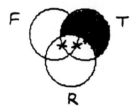

2. Invalid

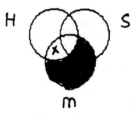

5. Valid

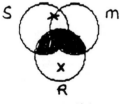

6. Invalid

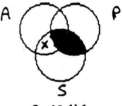

9. Valid

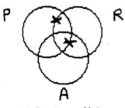

10. Invalid

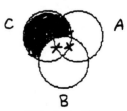

13. Invalid

66

Exercise 15.2 continued

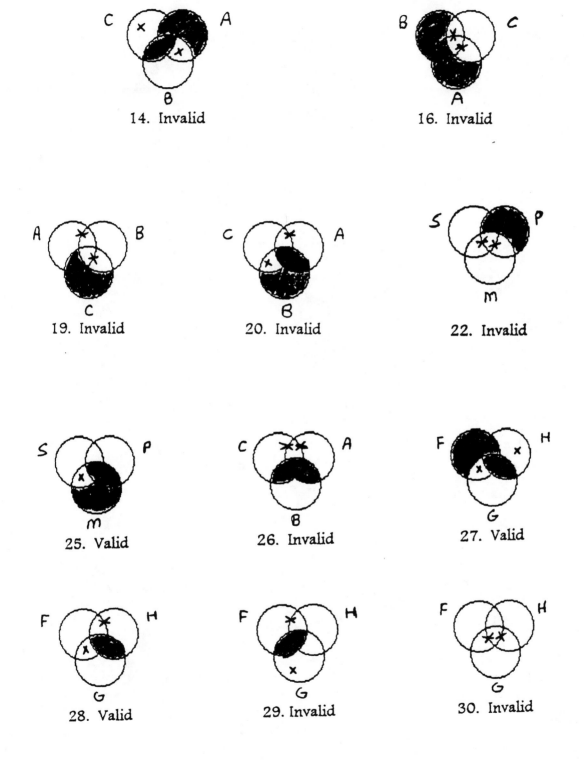

14. Invalid

16. Invalid

19. Invalid

20. Invalid

22. Invalid

25. Valid

26. Invalid

27. Valid

28. Valid

29. Invalid

30. Invalid

Exercise 15.3

2. Invalid

3. Valid

5. Valid

6. Invalid

7. Invalid

9. Valid

10. Invalid

13. Invalid

14. Invalid

15. Invalid

19. Invalid

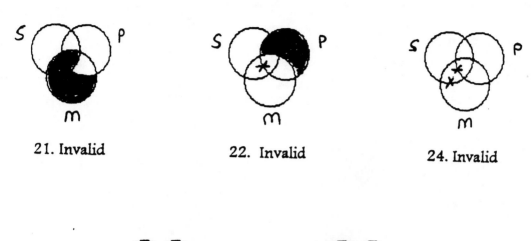

21. Invalid

22. Invalid

24. Invalid

26. Invalid

27. Valid

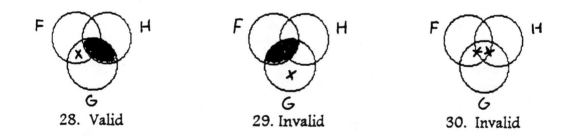

28. Valid

29. Invalid

30. Invalid

Exercise 15.4

3. All energetic persons are happy persons.
   All athletes are energetic persons.
   So, all athletes are happy persons.

   All athletes are happy persons.
   No students of psychology are happy persons.
   So, no athletes are students of psychology.

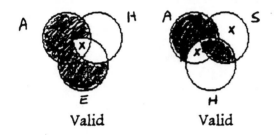

Valid   Valid

5. All A are B
   Some C are A
   So, some C are B

   Some C are B
   All C are D
   So, some D are B

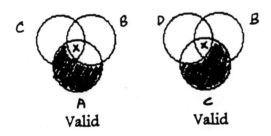

Valid   Valid

8. All A are B
   Some C are A
   So, some C are B

   Some C are B
   All C are D
   So, some D are B.

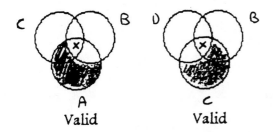

Valid   Valid

9. All A are B
   No C are B
   So, No C are A

   No C are A
   All D are C
   So, no D are A

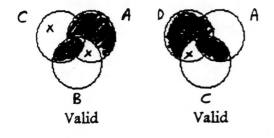

Valid   Valid

10. No B are D
    All A are B
    So, no A are D

    No A are D
    Some C are not D
    So, no A are C

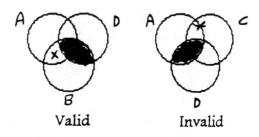

Valid   Invalid

Exercise 15.5

2.   Pat is a woman.
3.   Susan sympathizes with Kramer.
6.   Joe's truck is not gas-powered.
8.   Many adults use caffeine.
10.   Gaining a superficial view of the world is a waste of time.
12.   All mammals have hair.

Exercise 15.6

2.   All computers are electronic. Some guitars are electronic. So, some guitars are computers.
5.   No cars are trucks. No trucks are made by Rolls Royce. So, no vehicles made by Rolls Royce are cars.
6.   Some mammals are not feline. All cats are mammals. So, some cats are not feline.
9.   Some cats are pets. All wild lions are cats. So, some wild lions are pets.
10. All cats are mammals. Some pets are not cats. Therefore, no pets are mammals.
13. Some musical instruments are electronic devices. Some electronic devices are calculators. So, some musical instruments are calculators.

Supplementary Exercise

If you would like more practice with syllogisms, here are some additional exercises.

   1. Construct a syllogism of the form AEE-2 using the following three terms:
        rats, mammals, cute things.

   2. Construct a syllogism of the form EAE-1 using the following three terms:
        aardvarks, brown things, ugly things

   3. Construct a syllogism of the form AII-1 using the following three terms:
        kangaroos, brown things, funny-looking things

   4. Construct a syllogism of the form OAO-3 using the following three terms:
        koalas, cute things, little things

   5. Construct a syllogism of the form AEO-2 using the following three terms:
        wet things, round things, green things

   6. Construct a syllogism of the form EAO-2 using the following three terms:
        illie pies, purple things, weird things

   7. Construct a syllogism of the form EAO-1 using the following three terms:
        heavy-metal rockers, wild things, rebellious persons

   8. Construct a syllogism of the form AEO-4 using the following three terms:
        monks, spirituals persons, quiet persons.

Chapter 16

Chapters 2 through 4 introduced our first formal language, TL, the language for *truth-functional* logic. This chapter introduces a second formal language, QL, the language for *quantificational* logic. The language QL will allow us to translate and formalize the categorical sentences studied in Chapters 14-15. In general, QL handles sentences that contain quantifiers such as *all* and *some* and it also handles singular sentences such as *Herman Snodgrass ate three corndogs for lunch.*.

This chapter and the next will prepare you for a second system of natural deduction-- natural deduction for quantified arguments. Thus, in Chapter 18, you will learn the natural deduction system that accompanies QL--the system QD.

Answers

Exercise 16.1

2. Hw v Sp   Abbreviations: Hx: x eats a hamburger. Sx: x eats spinach. w: Wimpy p: Popeye
3. Ts & Cs   Abbreviations: Tx: x has 21 decks. Cx: x has a compliment of 72 officers and
                428 enlisted crew members. s: Starship *Enterprise*.
6. ~Sj v ~Sk   Abbreviations: Sx: x swims. j: Jerry k: Kramer

Exercise 16.2

2. (∃x)Rx   Abbreviation: Rx: x is right
3. (∃x)~Rx
6. ~(∃x)~Rx
8. (∃x) (Fx & ~Hx)   Abbreviations: Fx:x is a fish   Hx: x is hungry
10. (∃x) (Ox & Cx)   Abbreviations: Ox: x is an opossum   Cx: x is cuddly
12. ~ (x) (Ox ⊃ Cx)
13. ~(∃x) (Rx & ~Cx)   Abbreviations: Rx: x is a rat   Cx: x is a cute animal

Exercise 16.3

2. ~(∃x) (Ox & Ix)   Abbreviations: Ox: x is an old book Ix: x is interesting
Of course, this may also be symbolized: (x) (Ox ⊃ ~Ix)
3. (x) (Px ⊃ Hx) ⊃ Sn Abbreviations: Px: x is a person. Hx: x is happy. Sx: x is sad. n:
                    Newman
6. (x) (Mx v ~Mx)   Abbreviation: Mx: x is material
7. (x) (Hx & Ix) ⊃ Ax Abbreviations: Hx: x harms someone Ix: x is intentional Ax: x is against
                    the law
8. (∃x) (Px & Sx) ⊃ (∃x) (Px & Lx)   Abbreviations: Px: x is a person Sx: x slips on a banana
                    peel   Lx: x laughs
10. ~(∃x) Fx   Abbreviations: Fx: x is a flying dragon
12. (∃x)Nx   Abbreviations: Nx: x is a nondetectable strain of AIDs.
13. (x)(Fx⊃Bx)   Abbreviations: Fx: x is free Bx: x is brave
15. ~(x)(Kx⊃Bx)   Abbreviations: Kx: x is a klingon   Bx: x is bad

18.(x)(Hx⊃Tx)   Abbreviations: Hx: x is a hot dog with peanut butter Tx: x is a tasty treat.

19.(∃x)[Px&(Bx v Px)] Abbreviations: Px:x is a person. Bx:x is too busy. Px: x is too preoccupied.

20.~(x)(Lx ⊃ Gx) Abbreviations:Lx: x is a long book   Gx: x is a good book

21. (x) (Dx ⊃ Hx) Abbreviations: Dx: x has a diet... Hx: x has a high...

22.(x) (Lx ⊃ Hx) Abbreviations: Lx: x likes the Marx Brothers Hx: x has good taste...

24.(x) (Tx & Px) Abbreviations:Tx: x is temporary Px: x is perishable

25.(x)(Tx & Px)⊃L Abbreviation: L: Life has no meaning.

27.(x)(Px & ~Dx)⊃ Hx] & (x)(Dx ⊃ ~Hx) Abbreviation:Dx: x is a demonstrator

28.(x)[(Px & Ax)⊃ Fx] Abbreviations:Ax: x was in attendance Fx: x is a fan of the original Star Trek.

29.(x)(Px ⊃ Rx) Abbreviations: Px: x was partying Rx: x is a rocker.

30.(x)(Lx ⊃ Sx)Abbreviations: Lx: x likes Lost in Space. Sx: x is a Star Trek fan.

Chapter 17

This chapter has two main aims:

1. to introduce dyadic predication
2. to introduce overlapping quantifiers

Although some of this material is difficult, I believe that you will deepen your understanding of language and grammar when you study this logical theory. Natural deduction proofs with overlapping quantifiers and dyadic predicates are not covered until Chapter 21. This chapter is therefore necessary background material for Chapter 21.

Answers

Exercise 17.1

2. ~ (∃x) (Lx & Aax)   Abbreviations: Axy: x likes y Lx: x is a liberal a: Archie B.
   or (x) (Lx ⊃ ~Aax)
3. ~(x) (Lx ⊃ Aax)
4. Rww   Abbreviations: Rxy: x respects y w: Wimpy
7. (x) (Ex ⊃ Lxn)   Abbreviations:Ex: x is an elephant. Lxy: x is larger than y. n: Nathan
8. (x) (Hx ⊃ Llx)   Abbreviations:Hx: x is a horse Lxy: x likes y. l: Lorraine
9. (∃x) (Px & Kxk)   Abbreviations:Px: x is a person. Kxy: x knows y. k: Katie
11. (x) (Cx ⊃ Lex)   Abbreviations:e: Elmer Cx: x contains caffeine Lxy: x likes y.
14. (x) (Px ⊃ Lxr) & ~(∃x) (Px & Lxn) Abbreviations:Lxy: x loves y n: Ned r: Raymond
15. (x) (Px ⊃ Lxr) ⊃ (∃x) (Px & Lxg) Abbreviations: g: George
16. (∃x) (Px & Lxj) ⊃ Hj   Abbreviations: j: Jan. Hx: x is pleased. Px: x is a person. Lxy: x likes y
17. ~(∃x) (Px & Lxj) ⊃ Sj   Abbreviations: Sx: x is sad
18. (x) (Px ⊃ Lxx) ⊃ (x) (Px ⊃ Sx)   Abbreviations: Sx: x is selfish Px: x is a person
19. ~(∃x) Cxj   Abbreviations: Cxy: x cares about y j: Joe
20. ~(∃x) Cjx
21. ~(x) (Px ⊃ Lxr)

Exercise 17.2

2. (∃x) (y) Ryx   (Domain: persons)
3. (∃x) (y) Kxy
4. (x) (∃y) Rxy
8. (x) (∃y) (Lxy ⊃ ~Bx)
9. (x) [(Lx & Rxx) ⊃ Hx]
11. (x) (Ax ⊃ Tpx) & (x) (Ex ⊃ Spx)
13. ~(x) (Hx ⊃ Px)
14. (x) (Hx ⊃ ~Px)
15. (∃x)~Px   (Domain: events Px: x has a purpose)
16. (x) (∃y) Cyx

Exercise 17.3

2. ~(c = r)   Abbreviations: r: the richest person in the world. c: Charlie
or:~(x)[~(x = c) ⊃ Rcx]   Domain:persons.Rxy:x is richer than y.
3.(∃x)(∃y)[(Rx&Ry)&~(x=y)]   (Rx: x is Rita's job)
4.(x)(y)[(Px&Py) ⊃ (x=y)]
8.(∃x)(y)[~(y=x) ⊃ Oxy]   Domain: persons
      Universal domain: (∃x)(y){Px&[(Py&~(y=x)) ⊃Oxy]}
10.(∃x)(∃y)(z){[(Dx&Dy)& ~(x=y)]&[Dz ⊃ (z=x v z=y)]}
11.~(∃x)Dx
12.(x)[~(x=g) ⊃ Ggx]   Abbreviations: g: God. Gxy:x is greater than y.
13.(∃x) (∃y)(z) { [(Dx & Dy) & ~(x=y)] & [Dz ⊃ ((z = x) v (z = y))] }
14.Dr&(x){[Dx& ~(x=r)] ⊃ Brx}
17.(x){[Px& ~(x=c)] ⊃ Hx}&~Hc
18.Tr&(x)[(Tx& ~(x=r)) ⊃ Orx]
19.(~Hp&~Hc)&(x){[Px&(~(x=p)& ~(x=c))] ⊃ Hx}
21. (Kj & Ke) & (x) {[Sx & ~ (x = e) & ~ (x = j)] ⊃ ~Kx}
Sx: x is a solar system body
Kx: x is known to have erupting volcanoes
22. (x) (y) (z) (z') {[Dx & Dy & Dz & Dz' & ~(x = y) & ~ (y = z) & ~ (x = z)] ⊃ [(z' = x) v (z' = y) v (z' = z)]} Note: parentheses on the quadruple conjunction and triple conjunction have been left off to avoid excessive clutter.
23. (x) [Mxr ⊃ (x = m)]
25. (x) {[Sx & ~(x = s)] ⊃ Lx}   Sx: x is a star   Lx: x lies outside our solar system   s: the Sun
26. (x){[Gx & ~ (x = v)] ⊃ Lvx} Gx: x is a galaxy v: Virgo A Lxy: x is known to be larger than y
27. ~(∃x) (Px & Lxj)

Exercise 17.4

2.   a.   asym., irref., trans.
     c.   asym., irref., trans.
     d.   nonsym., nonref., nontrans.
     f.   nonsym., nonref., nontrans.
     g.   asym., irref., trans.
     h.   nonsym., irref., nontrans.
     i.   nonsym., ref., trans.
     k.   nonsym., ref., nontrans.

Chapter 18

The proofs in this chapter require no overlapping quantifiers and no dyadic predicates. The only quantificational logic presupposed is therefore Chapter 16. Of course, the truth-functional inference and replacement rules are also used in this unit.

If you covered Chapter 15, you might wish to compare the Venn diagram method of that chapter with the natural deduction method of this chapter. The two methods differ in some respects and overlap in other respects. It never hurts to compare and contrast things. We learn about a thing when we contrast it with other things.

Answers

Exercise 18.1

2.   3.   EI 2
      4.   Simp 3
      5.   UI 1
      6.   MP 4, 5
      7.   Simp 3
      8.   DS 6, 7
      9.   Conj 4, 8
      10.   EG 9

4.   4.   MT 1, 2
      5.   MP 3, 4

Exercise 18.2

```
(2)1.(x)(Hx⊃Jx)
2.(∃x)(Hx)/(∃x)(Jx)
3.Hv                    EI 2
4.Hv⊃Jv                 UI 1
5.Jv                    MP 3, 4
6.(∃x)(Jx)             EG 5

(3)1.(x)(Sx⊃Gx)
2.Sa /Ga
3.Sa ⊃ Ga      UI 1
4.Ga           MP 2, 3

(6)1.(∃x)(Fx&~Mx)
2.(x)(Fx⊃Hx)  /(∃x)(Hx&~Mx)
3.Fv & ~Mv     EI 1
4.Fv ⊃ Hv      UI 2
5.Fv           Simp 3
6.~Mv          Simp 3
```

```
7.Hv              MP 4, 5
8.Hv&~Mv          Conj 6, 7
9.(∃x)(Hx&~Mx)    EG 8


(8)1.(x)(Hx&Sx)
2.(∃x)(Hx)⊃(∃x)(Gx)/(∃x)(Sx)&(∃x)(Gx)
3.Ha&Sa               UI 1
4.Ha                  Simp 3
5.Sa                  Simp 3
6.(∃x)Sx              EG 5
7.(∃x)Hx              EG 4
8.(∃x)Gx              MP 2, 7
9.(∃x)(Sx)&(∃x)(Gx) Conj 6,8


(11)1.(x)(Ax⊃Bx)
2.(x)(Ax)/x)(Bx)
3.Au⊃Bu    UI 1
4.Au              UI 2
5.Bu              MP 3, 4
6.(x)(Bx)         UG 5


(13)1.(x)(Sx)⊃(x)(Gx)
2.~(x)(Gx)  /  ~(x)(Sx)
3.~(x)(Sx)        MT 1, 2
```

USEFUL VOCABULARY FOR LOGIC STUDENTS

MODAL

DOUBLE ARROW

FINAL COLUMN

EXISENTIAL INSTANTIATION

INDUCTIVE ARGUMENT

NECESSARY TRUTH

THEOREM

AMPERSAND

```
(14)1. (∃x)(Bx) ⊃(x)(Hx⊃Gx)
2.  Bb&Hb / Gb
3.Bb                    Simp 2
4.(∃x)(Bx)              EG 3
5.(x)(Hx⊃Gx)            MP 1, 4
6.Hb⊃Gb                 UI 5
7.Hb                    Simp 2
8.Gb                    MP 6, 7


(15)1.(x)(Fx⊃Sx)
2.Fa&Fb / Sa & Sb
3.Fa⊃Sa                 UI 1
4.Fa                    Simp 2
5.Sa                    MP 3, 4
6.Fb⊃Sb                 UI 1
7.Fb                    Simp 2
8.Sb                    MP 6, 7
9.Sa&Sb                 Conj 5, 8
```

Exercise 18.3

```
(2)1.(x)(Axv~Bx)
2.Hs / (∃x)[(Hx&Ax)v(Hx&~Bx)]
3.Asv~Bs                 UI 1
4.Hs&(Asv~Bs)            Conj 3, 2
5.(Hs&As)v(Hs&~Bs)       Dist 4
6.(∃x)[(Hx&Ax)v(Hx&~Bx)] EG 5


(3)1.(∃x)(Jx)⊃(x)(HxvSx)
2.(x)(Fx v~Sx)
3.(x)(Jx) /(∃x)(Hx v Fx)
4.Ja                    UI 3
5.(∃x)(Jx)              EG 4
6.(x)(HxvSx)            MP 1, 5
7.HavSa                 UI 6
8.Fav~Sa                UI 2
9.~~HavSa               DNeg 7
10.~Ha⊃Sa               Imp 9
11.~SavFa               Comm 8
12.Sa⊃Fa                Imp 11
13.~Ha⊃Fa               HS 10, 12
14.~~HavFa              Imp 13
15. HavFa               DNeg 14
16.(∃x)(HxvFx)          EG 13
```

(7).1.(x)[Hx⊃(SxvGx)]
2.(∃y)(~Sy&~Gy) /(∃x)(~ Hx)
3.~Sv&~Gv   EI 2
4.Hv⊃(Sv v Gv)   UI 1
5.~(Sv v Gv)   DM 3
6.~Hv   MT 4, 5
7.(∃x)~Hx   EG 6

Exercise 18.4

(3)1.(x)(Hx)⊃(∃x)(Sx)
2.(x)(~Sx) /(∃x)(~Hx)
3.~(∃x)(Sx)   QE 2
4.~(x)(Hx)   MT 1, 3
5.(∃x)(~Hx)   QE 4

(4)1.(∃x)(~Hx)v(∃x)(~Sx)
2.(x)(Sx) /~(x)(Hx)
3.~(∃x)~Sx   QE 2
4. (∃x)(~Sx)v(∃x)(~Hx) Comm 1
5.(∃x)~Hx   DS 3, 4
6.~(x)(Hx)   QE 5

(6)1.(∃x)(Gx)v(∃x)(Hx&Sx)
2.~(∃x)(Hx)/(∃x)(Gx)
3.(x)(~Hx)   QE 2
4.~Hu   UI 3
5.~Huv~Su   Add 4
6.~(Hu&Su)   DM 5
7.(x)~(Hx&Sx)   UG 6
8.~(∃x)(Hx&Sx)   QE 7
9.(∃x)(Hx&Sx)v(∃x)(Gx)   Comm 1
10.(∃x)(Gx)   DS 8,9

(8)1.(x)(Jx)⊃(∃x)(~Sx)
2.~(x)(Sx)⊃(∃x)(~Hx) /(x)(Hx)⊃(∃x)(~Jx)
3.(∃x)(~Sx)⊃~(x)(Hx)   QE 2 (twice)
4.(x)(Jx)⊃~(x)(Hx)   HS 1,3
5.(x)(Jx)⊃(∃x)(~Hx)   QE 4
6.~(∃x)(~Hx)⊃~(x)(Jx)   Trans 5
7.(x)(Hx)⊃(∃x)(~Jx)   QE 6 twice

(11) 1. (x)[(Px&Qx)⊃Rx]
2. ~(x)(Px⊃Rx)   /(∃x)(~Qx)
3. (∃x)~(Px⊃Rx)   QE 2
4. ~(Pv⊃Rv)   EI 3
5. ~(~PvvRv)   Imp 4
6. ~~Pv&~Rv   DM 5
7. Pv&~Rv   DNeg 6
8. (Pv&Qv)⊃Rv   UI 1
9. ~Rv   Simp 7
10. ~(Pv&Qv)   MT 8, 9
11. ~Pvv~Qv   DM 10
12. Pv   Simp 7
13. ~~Pv   DNeg 12
14. ~Qv   DS 11, 13
15. (∃x)(~Qx)   EG 14

(12) 1. (∃x)(~Hx)⊃(x)(Ax⊃Bx)
2. ~(x)(HxvBx)   /(∃x)(~Ax)
3. (∃x)~(HxvBx)   QE 2
4. ~(Hv v Bv)   EI 3
5. ~Hv&~Bv   DM 4
6. ~Hv   Simp 5
7. (∃x)(~Hx)   EG 6
8. (x)(Ax⊃Bx)   MP 1, 7
9. Av⊃Bv   UI 8
10. ~Bv   Simp 5
11. ~Av   MT 9, 10
12. (∃x)(~Ax)   EG 11

(13) 1. ~(x)(Ax⊃Bx)
2. (x)(Dx⊃Bx)   /(∃x)(Ax&~Dx)
3. (∃x)~(Ax⊃Bx)   QE 1
4. ~(Av⊃Bv)   EI 3
5. ~(~Av v Bv)   Imp 4
6. ~~Av&~Bv   DM 5
7. Av&~Bv   DNeg 6
8. ~Bv   Simp 7
9. Dv⊃Bv   UI 2
10. ~Dv   MT 8,9
11. Av   Simp 7
12. Av&~Dv   Conj 10,11
13. (∃x)(Ax&~Dx)   EG 12

Exercise 18.5

2. 1. ~(∃x)(Cx&Rx)
    2. <u>(∃x)(Px&Rx)</u>/(∃x)(Px&~Cx)
    3. (x)~(Cx&Rx)   QE 1
    4. Pv&Rv   EI 2
    5. ~(Cv&Rv)   UI 3
    6. ~Cvv~Rv   DM 5
    7. Pv   Simp 4
    8. Rv   Simp 4
    9. ~Rv v ~Cv   Comm 6
    10. ~~Rv   DNeg 8
    11. ~Cv   DS 9, 10
    12. Pv&~Cv   Conj 7, 11
    13. (∃x)(Px&~Cx)   EG 12

4. 1. (x)(Dx⊃Mx)
    2. ~(∃x)(Ax&Mx)
    3. <u>(x)(Ax⊃~Dx)⊃(x)(Ax⊃~Bx)</u>/~(∃x)(Ax&Bx)
    4. (x)~(Ax&Mx)   QE 2
    5. ~(Au&Mu)   UI 4
    6. ~Auv~Mu   DM 5
    7. Au⊃~Mu   Imp 6
    8. Du⊃Mu   UI 1
    9. ~Mu⊃~Du   Trans 8
    10. Au⊃~Du   HS 7, 9
    11. (x)(Ax⊃~Dx)   UG 10
    12. (x)(Ax⊃~Bx)   MP 3, 11
    13. Au⊃~Bu   UI 12
    14. ~Au v ~Bu   Imp 13
    15. ~(Au&Bu)   DM 14
    16. (x)~(Ax&Bx)   UG 15
    17. ~(∃x)(Ax&Bx)   QE 16

6. 1. Pc
    2. <u>(x)(Px⊃Hx)</u>/Hc (note: Hx: x is a person)
    3. Pc⊃Hc   UI 2
    4. Hc   MP 1, 3

7. 1. (x)[(CxvDx)⊃Rx](note:Rx:x reasons,learns,loves)
    2. <u>(x)[Rx⊃(Px&Gx)]</u>/(x)[(CxvDx)⊃(Px&Gx)]
    3. (CuvDu)⊃Ru   UI 1
    4. Ru⊃(Pu&Gu)   UI 2
    5. (CuvDu)⊃(Pu&Gu)   HS 3, 4
    6. (x)[(CxvDx)⊃(Px&Gx)]   UG 5

82

```
10.  1.    (x)(Dx⊃Sx)
     2.    (x)(~Dx⊃~Sx)
     3.    ~Sb/~Db
     4.    Db⊃Sb              UI 1
     5.    ~Db                MT 3, 4
11.  1.    ~(∃x)(Cx&Rx)
     2.    (∃x)(Mx&Cx)/(∃x)(Mx&~Rx)
     3.    Mv&Cv              EI 2
     4.    (x)~(Cx&Rx)        QE 1
     5.    ~(Cv&Rv)           UI 4
     6.    ~Cv∨~Rv            Dm 5
     7.    Cv⊃~Rv             Imp 6
     8.    Cv                 Simp 3
     9.    ~Rv                MP 7, 8
     10.   Mv                 Simp 3
     11.   Mv&~Rv             Conj 9, 10
     12.   (∃x)(Mx&~Rx)       EG 11

13.  1.    (x)(Fx⊃Rx)
     2.    Sd⊃Rd
     3.    Sd&~Fd/(∃x)(Rx&~Fx)
     4.    Sd                 Simp 3
     5.    Rd                 MP 2, 4
     6.    ~Fd                Simp 3
     7.    Rd&~Fd             Conj 5, 6
     8.    (∃x)(Rx&~Fx)       EG 7

16.  1.    As⊃(x)(Ax)
     2.    As/Ag
     3.    (x)(Ax)            MP 1, 2
     4.    Ag                 UI 3

17.  1.    R⊃N
     2.    G⊃R
     3.    G/N
     4.    R                  MP 2, 3
     5.    N                  MP 1, 4

18.  1.    (x)(Cx)∨(x)(Mx)
     2.    (x)(Mx)⊃~L
     3.    ~G ⊃~(x)(Cx)/~G ⊃ ~L
     4.        ~G             AP
     5.        ~(x)(Cx)       MP 3, 4
     6.        (x)(Mx)        DS 1, 5
     7.        ~L             MP 2, 6
     8.    ~G ⊃ ~L            CP 4-7
```

(Abbreviations:G: God exists. L: Life has transcendent meaning.)

(20) 1. (x)(Ax⊃Mx)
2. (x)(Mx⊃Wx)
3. Aa
4. Wa⊃~Ca / ~Ca
5. Aa⊃Ma   UI 1
6. Ma⊃Wa   UI 2
7. Aa⊃Wa   HS 5,6
8. Aa⊃~Ca   HS 4,7
9. ~Ca   MP 3,8

(21) 1. (x)(Ex⊃Px)
2. ~(∃x)(Px&Dx) /(x)(Ex⊃~Dx)
3. (x)~(Px&Dx)   QE 2
4. ~(Pu&Du)   UI 3
5. Eu⊃Pu   UI 1
6. ~Pu v ~Du   DM 4
7. Pu⊃~Du   Imp 6
8. Eu⊃~Du   HS 5,7
9. (x)(Ex⊃~Dx)   UG 8

(22) 1. (x)[Ex⊃(Gx&Hx)]
2. (∃x)(Ex&Cx)/(∃x)[Cx&(Gx&Hx)]
3. Ev &Cv   EI 2
4. Ev   Simp 3
5. Cv   Simp 3
6. Ev⊃(Gv&Hv)   UI 1
7. Gv&Hv   MP 4,6
8. Cv&(Gv&Hv)   Conj 5,7
9. (∃x)[Cx&(Gx&Hx)]   EG 8

(23) . 1. (∃x)(Cx&Ex)
2. (x)[(Ex&Cx)⊃Rx]
3. (∃x)(Cx&Rx)⊃Wj   /Wj
4. Cv&Ev   EI 1
5. Ev& Cv   Comm 4
6. (Ev&Cv)⊃Rv   UI 2
7. Rv   MP 5,6
8. Cv   Simp 4
9. Cv&Rv   Conj 7,8
10. (∃x)(Cx&Rx)   EG 9
11. Wj   MP 3,10

Chapter 19

This chapter has two main aims:

1. To specify the formal semantics for quantificational logic.
2. To introduce two methods of proving invalidity:
      a. refutation by logical analogy
      b. the method of truth-functional expansions (for monadic predicate arguments).

Incidentally, if you want more practice with the "refutation by logical analogy" problems of Exercise 19.2, additional problems can be found in Chapter 15, Exercise 15.6 (page 328).

Answers

Exercise 19.1

2.   Domain: motor vehicles
        Ax: x is a Ford
        Bx: x is a car

3.   Domain: professional athletes
        Ax: x is a professional baseball player
        Bx: x is healthy

6.   Domain: food
        Ax: x is a hamburger
        Bx: x is fattening

7.   Ax: x is a cat
        Bx: x is a mammal
        Domain: animals

8.   Ax: x is a cat
        Bx: x is a mammal
        Domain: animals

9.   Ax: x is a cat
        Bx: x is a dog
        Domain: animals

12.   Fx: x bought a losing lottery ticket
        Gx: x bought a winning lottery ticket
        b: Bill Clinton

13.   Ax: x is a animal
        Bx: x is a plant
        Cs: x contains chlorophyll
        Domain: living things

14.   Ax: x is a terrier
        Gx: x is a cat
        Hx: x is a dog
        Domain: animals

15.   Ax: x is an angel
        Bx: x is an immaterial being
        Domain: universal domain

Part II.

2. No cats are reptiles. No reptiles are mammals. So, no cats are mammals.

3. All cats are mammals. Some mammals are bats. So, some cats are bats.

4. All rats are mammals. All humans are mammals. So, all rats are humans.

7. All dogs are mammals. All mammals have hearts. So, all beings with hearts are dogs.

Exercise 19.2

2.   (Aa&Ja)v(Ab&Jb)
        (Aa&~Ja)v(Ab&~Jb)
        Aa: T   Ab: F
        Ja: T   Jb: F

3.   Ha&Hb
        (Ha&Ga)v(Hb&Gb)
        Ha: T   Ga: F
        Hb: T   Gb: F

5.   Aa ⊃ Ba
        ~Aa ⊃ ~Ba

        Aa: F Ba: T

6.   Aa ⊃ Ba
        Ba
        Aa

        Aa: F Ba: T

8. <u>(Aa v Ba) & (Ab v Bb)</u>

(Aa & Ab) v (Ba & Bb)

Aa: T Ba: F Ab: F Bb: T

9.   Aa ⊃ Ba
      <u>Ba</u>
      Aa
Aa: F Ba: T

Chapter 20

This chapter focuses on indirect and conditional quantificational proofs. Of course, this chapter presupposes the material on conditional and indirect proofs in Chapters 9 and 11. You might say that this chapter is "conditional upon" the material in those two chapters.

Answers

Exercise 20.1

```
(2)1.(x)(Hx⊃Sx)
2.(x)(Sx⊃Gx)  /(x)[Hx⊃(Sx&Gx)]
   3.        ┃ ~(x)[Hx⊃(Sx&Gx)]      AP
   4.        ┃ (∃x)~[Hx⊃(Sx&Gx)]     QE 3
   5.        ┃ ~[Hv⊃(Sv&Gv)]         EI 4
   6.        ┃ ~[~Hv v(Sv&Gv)]       Imp 5
   7.        ┃ ~~[~~Hv&~(Sv&Gv)]     DM 6
   8.        ┃ Hv &~(Sv&Gv)          DNeg 7 (twice)
   9.        ┃ Hv                     Simp 8
  10.        ┃ ~(Sv&Gv)              Simp 8
  11.        ┃ ~Sv v ~Gv             DM 10
  12.        ┃ Hv⊃Sv                  UI 1
  13.        ┃ Sv⊃Gv                  UI 2
  14.        ┃ Hv⊃Gv                  HS 12, 13
  15.        ┃ Gv                     MP 14, 9
  16.        ┃ ~~Gv                   DNeg 15
  17.        ┃ ~Gv v ~Sv             Comm 11
  18.        ┃ ~Sv                    DS 16, 17
  19.        ┃ ~Hv                    MT 12, 18
  20.        ┃ Hv & ~Hv              Conj 9,19
 21.(x)[Hx⊃(Sx&Gx)]                   IP 3-20
```

```
(4)1.(x)[Ax⊃(Bx v Cx)]/(∃x)(Ax)⊃(∃x)(Bx v Cx)
   2. ┃(∃x)(Ax)           AP
   3. ┃Av                 EI 2
   4. ┃Av⊃(Bv v Cv)       UI 1
   5. ┃Bv v Cv            MP 3, 4
   6. ┃(∃x)(Bx v Cx)      EG 5
 7.(∃x)(Ax)⊃(∃x)(BxvCx)   CP 2-6
```

```
(8)1.(x)~(Hx&~Bx)
2.(x)~(Bx&~Gx)  /(x)~(Hx&~Gx)
   3. ┃~(x)~(Hx&~Gx)      AP
   4. ┃(∃x)(Hx&~Gx)       QE 3
   5. ┃Hv&~Gv             EI 4
```

```
6.  | ~(Hv&~Bv)           UI 1
7.  | ~Hv v ~~Bv          DM 6
8.  | ~Hv v Bv            DNeg 7
9.  | ~(Bv&~Gv)           UI 2
10. | ~Bv v ~~Gv          DM 9
11. | ~Bv v Gv            DNeg 10
12. | ~Gv                 Simp 5
13. | Gv v ~Bv            Comm 11
14. | ~Bv                 DS 12, 13
15. | Bv v ~Hv            Comm 8
16. | ~Hv                 DS 14, 15
17. | Hv                  Simp 5
18. | Hv & ~Hv            Conj 16, 17
19. (x)~(Hx&~Gx)          IP 3-18


(9)1.(x)[(Fx v Gx)⊃Px]
2.(∃x)(~Fx v Sx)⊃(x)(Rx)  /  (x)(Px)v(x)(Rx)
3.      | ~[(x)(Px)v(x)(Rx)]    AP
4.      | ~(x)(Px)&~(x)(Rx)     DM 3
5.      | ~(x)(Px)              Simp 4
6.      | ~(x)(Rx)              Simp 4
7.      | (∃x)(~Px)            QE 5
8.      | (∃x)(~Rx)            QE 6
9.      | ~Pv                   EI 7
10.     | ~Rv'                  EI 8
11.     | (Fv v Gv)⊃Pv          UI 1
12.     | ~(Fv v Gv)            MT 9, 11
13.     | ~Fv&~Gv               DM 12
14.     | ~Fv                   Simp 13
15.     | ~Fv v Sv              Add 14
16.     | (∃x)(~FxvSx)         EG 15
17.     | (x)(Rx)               MP 2, 16
18.     | Rv'                   UI 17
19.     | Rv' & ~Rv'            Conj 18, 10
20. (x)(Px)v(x)(Rx)            IP 3-19


(12)1.(x)(Jx⊃Px)
2.(x)(Hx⊃Mx) /(∃x)(Jx v Hx)⊃(∃x)(Px v Mx)
3.    | (∃x)(Jx v Hx)           AP
4.    | Jv v Hv                  EI 3
5.    | Jv⊃Pv                    UI 1
6.    | Hv⊃Mv                    UI 2
7.    | Pv v Mv                  CD 4, 5, 6
8.    | (∃x)(Px v Mx)           EG 7
9.(∃x)(Jx v Hx)⊃(∃x)(Px v Mx)   CP 3-8
```

(14) 1. (∃x)(Qx)⊃(x)(Sx)
2. Qa⊃~Sa /~Qa
```
3.          | ~~Qa              AP
4.          | Qa                DNeg 3
5.          | ~Sa               MP 2, 4
6.          | (∃x)(Qx)          EG 4
7.          | (x)(Sx)           MP 1, 6
8.          | Sa                UI 7
9.          | Sa & ~Sa          Conj 5,8
10. ~Qa                         IP 3-9
```

(16) 1. (∃x)(Hx)⊃(∃x)(Sx&Fx)
2. ~(∃x)(Fx) / (x)(~Hx)
```
3.          | ~(x)(~Hx)         AP
4.          | (∃x)(Hx)          QE 3
5.          | (∃x)(Sx&Fx)       MP 1, 4
6.          | Sv&Fv             EI 5
7.          | (x)(~Fx)          QE 2
8.          | ~Fv               UI 7
9.          | Fv                Simp 6
10.         | Fv & ~Fv          Conj 8,9
11. (x)(~Hx)                    IP 3-10
```

(17) 1. (∃x)(Sx)⊃(∃x)(Hx&Jx)
2. (x)(Fx⊃Sx) / (∃x)(Fx)⊃(∃x)(Hx)
```
3.          | (∃x)(Fx)          AP
4.          | Fv                EI 3
5.          | Fv⊃Sv             UI 2
6.          | Sv                MP 4, 5
7.          | (∃x)(Sx)          EG 6
8.          | (∃x)(Hx&Jx)       MP 1, 7
9.          | Hv'&Jv'           EI 8
10.         | Hv'               Simp 9
11.         | (∃x)(Hx)          EG 10
12. (∃x)(Fx)⊃(∃x)(Hx)           CP 3-11
```

(18) 1. (∃x)(Px)⊃(∃x)(Qx&Sx)
2. (∃x)(Sx v Hx)⊃(x)(Gx) / (x)(Px⊃Gx)
```
3.          | ~(x)(Px⊃Gx)       AP
4.          | (∃x)~(Px⊃Gx)      QE 3
5.          | ~(Pv⊃Gv)          EI 4
6.          | ~(~Pv v Gv)       Imp 5
7.          | ~~Pv&~Gv          DM 6
8.          | Pv&~Gv            DNeg 7
9.          | ~Gv               Simp 8
10.         | Pv                Simp 8
```

90

```
11.        (∃x)(Px)              EG 10
12.        (∃x)(Qx&Sx)           MP 1, 11
13.        Qv'&Sv'               EI 12
14.        Sv'                   Simp 13
15.        Sv'vHv'               Add 14
16.        (∃x)(SxvHx)           EG 15
17.        (x)(Gx)               MP 2, 16
18.        Gv                    UI 17
19.        Gv & ~Gv              Conj 9,18
20.(x)(Px⊃Gx)                    IP 3-19


(19)1.(x)(Ax⊃Bx)
2.AbvAc /(∃x)Bx
3.Ab⊃Bb          UI 1
4.Ac⊃Bc          UI 1
5.Bb v Bc        CD 2,3,4
6.         ~(∃x)Bx        AP
7.         (x)~Bx         QE 6
8.         ~Bb            UI 7
9.         Bc             DS 5,8
10.        ~Bc            UI 7
11.        Bc & ~Bc       Conj 9,10
12. (∃x)Bx                IP 6-11
```

Exercise 20.2

```
2.    1.   ~[(x)(PxvQx)v(∃x)(~Pxv~Qx)]      AP
      2.   ~(x)(PxvQx)&~(∃x)(~Pxv~Qx)       DM 1
      3.   ~(x)(PxvQx)                       Simp 2
      4.   (∃x)~(PxvQx)                      QE 3
      5.   ~(PvvQv)                          EI 4
      6.   ~Pv&~Qv                           DM 5
      7.   ~(∃x)(~Pxv~Qx)                    Simp 2
      8.   (x)~(~Pxv~Qx)                     QE 7
      9.   ~(~Pvv~Qv)                        UI 8
      10.  ~~Pv&~~Qv                         DM 9
      11.  Pv&Qv                             DNeg 10
      12.  Pv                                Simp 11
      13.  ~Pv                               Simp 6
      14.  Pv & ~Pv                          Conj 12,13
      15.(x)(PxvQx)v(∃x)(~Pxv~Qx)            IP 1-14
```

3.   1. | ~(x)(Ax&Bx)   AP
     2. | (∃x)~(Ax&Bx)   QE 1
     3. | ~(Av & Bv)   EI 2
     4. | ~Av v~Bv   DM 3
     5. | (∃x)(~Axv~Bx)   EG 4
     6. ~(x)(Ax&Bx)⊃(∃x)(~Axv~Bx)   CP 1-5

6.   1. | ~~(∃x)(Px&~Px)   AP
     2. | (∃x)(Px&~Px)   DNeg 1
     3. | Pv&~Pv   EI 2
     4. ~(∃x)(Px&~Px)   IP 1-3

8.   1. | (x)(Fx)&(x)(Gx)   AP
     2. | (x)(Fx)   Simp 1
     3. | (x)(Gx)   Simp 1
     4. | Fu   UI 2
     5. | Gu   UI 3
     6. | Fu&Gu   Conj 4, 5
     7. | (x)(Fx&Gx)   UG 6
     8. [(x)(Fx)&(x)(Gx)]⊃(x)(Fx&Gx)   CP 1-7

9.   1. | ~[(x)(Sx)v(∃x)(~Sx)]   AP
     2. | ~(x)(Sx)&~(∃x)(~Sx)   DM 1
     3. | ~(x)(Sx)   Simp 2
     4. | ~(∃x)(~Sx)   Simp 2
     5. | (∃x)(~Sx)   QE 3
     6. | (x)(Sx)   QE 4
     7. | ~Sv   EI 5
     8. | Sv   UI 6
     9. | Sv & ~Sv   Conj 7,8
    10. (x)(Sx)v(∃x)(~Sx)   IP 1-9

11. 1. | (x)(Hx)   AP
     2. | Ha   UI 1
     3. | (∃x)(Hx)   EG 2
     4. (x)(Hx)⊃(∃x)(Hx)   CP 1-2

12. 1. | (x)(Sx⊃Px) AP
     2. | | (∃x)(Sx)   AP
     3. | | Sv   EI 2
     4. | | Sv⊃Pv   UI 1
     5. | | Pv   MP 3, 4
     6. | | (∃x)(Px)   EG 5
     7. | (∃x)(Sx)⊃(∃x)(Px)   CP 2-6
     8. (x)(Sx⊃Px)⊃[(∃x)(Sx)⊃(∃x)(Px)]   CP 1-7

Chapter 21

The focus in this chapter is proofs containing overlapping quantifiers. Of course, this material presupposes the material on relations and overlapping quantifiers in Chapter 17. You might say that chapters 17 and 21 "overlap."

Answers

Exercise 21.1 Part A

```
(2)1.(∃x)(y)(Sxy)  /  (y)(∃x)(Sxy)
2.(y)(Svy)        EI  1
3.Svu             UI  2
4.(∃x)(Sxu)       EG  3
5.(y)(∃x)(Sxy) UG  4
```

```
(3)1.(x)(Fx⊃Bx)
2.(x)(∃y)(Sxy v ~Bx)/(∃x)(∃y)(Sxy v ~Fx)
3.Fc⊃Bc                 UI  1
4.(∃y)(Scyv~Bc)         UI  2
5.Scv v ~Bc             EI  4
6.~Bc v Scv             Comm 5
7.Bc ⊃ Scv              Imp  6
8.Fc ⊃ Scv              HS 3, 7
9.~Fc v Scv             Imp  8
10.Scv v ~Fc            Comm 9
11.(∃y)(Scy v ~Fc)      EG 10
12.(∃x)(∃y)(Sxy v ~Fx)  EG 11
```

```
(5)1.(x)(∃y)(Sx&Py)  /(∃y)(∃x)(Sx&Py)
2.(∃y)(Sa&Py)          UI  1
3.(Sa&Pv)              EI  2
4.(∃x)(Sx&Pv)          EG  3
5.(∃y)(∃x)(Sx&Py)      EG  4
```

```
(6)1.(x)(∃y)(Jx v Ry)  /(∃y)(∃x)(Jx v Ry)
2.        ┃ ~(∃y)(∃x)(JxvRy)      AP
3.        ┃ (y)~(∃x)(JxvRy)       QE 2
4.        ┃ (y)(x)~(JxvRy)        QE 3
5.        ┃ (∃y)(JbvRy)           UI 1
6.        ┃ JbvRv                 EI 5
7.        ┃ (x)~(JxvRv)           UI 4
8.        ┃ ~(JbvRv)              UI 7
9.        ┃ JbvRv &~(JbvRv)       Conj 6,8
10.(∃y)(∃x)(JxvRy)                IP 2-9
```

93

(8)1.(∃x)(y)(Pxy⊃Sxy)
2.(x)(∃y)(~Sxy) /~(x)(y)(Pxy)
3. | ~~(x)(y)(Pxy)   AP
4. | (x)(y)(Pxy)   DNeg 3
5. | (y)(Pvy⊃Svy)   EI 1
6. | (∃y)(~Svy)   UI 2
7. | ~Svv'   EI 6
8. | Pvv'⊃Svv'   UI 5
9. | ~Pvv'   MT 7,8
10. | (∃y)(~Pvy)   EG 9
11. | (∃x)(∃y)(~Pxy)   EG 10
12. | (∃x)~(y)(Pxy)   QE 11
13. | ~(x)(y)(Pxy) .   QE 12
14. |~(x)(y)(Pxy) & ~~(x)(y)(Pxy)   Conj 3,13
15.~(x)(y)(Pxy)   IP 3-14

(9)1.(∃x)(y)Mxy/(x)(∃y)(Myx)
2.(y)Mvy   EI 1
3.Mvu   UI 2
4.(∃y)(Myu)   EG 3
5.(x)(∃y)(Myx)   UG 4

(10)1.(x)(y)[(Wx&Lxy)⊃Lya]
2.(x)(y)(Lxa⊃Lxy) /(x)(y)[(Wx&Lxy)⊃Lyx]
3.(y)[(Wu&Luy)⊃Lya] UI 1
4.(Wu&Luu')⊃Lu'a   UI 3
5.(y)(Lu'a⊃Lu'y)   UI 2
6.Lu'a⊃Lu'u   UI 5
7.(Wu&Luu')⊃Lu'u   HS 4,6
8.(y)(Wu&Luy)⊃Lyu   UG 7
9.(x)(y)[(Wx&Lxy)⊃Lyx]   UG 8

Exercise 21.1 Part B

(2) 1.(x)(y)(Mxy ⊃Lxy)
2.(x)(y)(Eyx ⊃Mxy)/ (x)(y)(Eyx ⊃ Lxy)
3.(y)(Muy ⊃Luy)   UI 1
4. Muu' ⊃Luu'   UI 3
5. (y)(Eyu ⊃ Muy)   UI 2
6. Eu'u ⊃ Muu'   UI 5
7. Eu'u ⊃ Luu'   HS 4,6
8. (y)(Eyu ⊃ Luy)   UG 7
9.(x)(y)(Eyx ⊃ Lxy)   UG 8

(3)1.(x)(∃y)(Cyx)
2.(∃y)(Cyu)⊃ C   / C
3.(∃y)(Cyu)   UI 1
4.C   MP 2,3

(Abbreviation:C: A Creator exists.Cxy: x causes y. u: the universe.)

(4)1.(x)(∃y)(Cyx)
2.G ⊃ (∃x)(y)~Cyx   / ~G
3.   ⎡~~G   AP
4.   ⎢G   DNeg 3
5.   ⎢(∃x)(y)~Cyx   MP 2,4
6.   ⎢(y)~Cyv   EI 5
7.   ⎢(∃y)(Cyv)   UI 1
8.   ⎢Cv'v   EI 7
9.   ⎢~Cv'v   UI 6
10.   ⎣Cv'v & ~Cv'v   Conj 8,9
11.~G   IP 3-10

(5)1. Tab
2.Tbe
3.(x)(y)(z){(Txy & Tyz)⊃ Txz}/ Tae
4. (y)(z){(Tay & Tyz)⊃ Taz}   UI 3
5. (z){(Tab & Tbz)⊃ Taz}   UI 4
6. (Tab & Tbe)⊃ Tae   UI 5
7. Tab & Tbe   Conj 1,2
8. Tae   MP 6,7

95

Chapter 22

The concept of *identity* was introduced in Chapter 17. In the present chapter, two natural deduction rules for proofs with identity are introduced. Many important and yet difficult philosophical arguments employ identity operators and concern the logic associated with the concept of identity. The logical techniques of this chapter are therefore useful for advanced work in philosophy.

Answers

Exercise 22.1

```
(4) 1.Ab⊃Bb
2.Rd⊃Sd
3.Ab&Rd
4.b = d    / Bd & Sb
5.Ad⊃Bd    IdB 1, 4
6.Ab       Simp 3
7.Ad       IdB 6, 4
8.Bd       MP 5, 7
9.Rb⊃Sb    IdB 2, 4
10.Rd      Simp 3
11.Rb      IdB 10, 4
12.Sb      MP 9, 11
13.Bd&Sb Conj 8, 12


(5)1.Hc⊃Kc
2. Md⊃Nd
3. Hc&Md
4. c=d / Kd & Nc
5.Hc          Simp 3
6.Kc          MP 1, 5
7.Kd          Id B 4, 6
8.Md          Simp 3
9.Nd          MP 2, 8
10.Nc         Id B 9, 4
11.Kd & Nc    Conj 7,10


(7)1.Ma⊃Wa
2. ~Wa
3. a = b / ~Mb
4. ~Ma        MT 1, 2
5. ~Mb        Id B 3,4
```

(8)1.Wa
2.(x)[Wx⊃(x = a)]
3.(∃x)(Wx&Bx) / Ba
4.Wv & Bv   EI 3
5.Wv   Simp 4
6.Wv ⊃ (v=a)   UI 2
7.v=a   MP 5,6
8.Bv   Simp 4
9.Ba   IdB 7,8

Chapter 23

Chapters 2, 3, and 4 introduced our first formal language, TL, the language for *truth-functional* logic. Chapters 16 and 17 introduced a second formal language, QL, the language for *quantificational* logic. This chapter introduces a third formal logical language, ML, the language for *modal* logic. This chapter will prepare you for a third system of natural deduction--natural deduction for modal arguments, which you will learn in the next chapter, Chapter 24.

Why study modal logic? Here is one reason: Many of the most interesting arguments in the history of philosophy are modal arguments--arguments built out of modal operators and the units to which these attach. There are deep and intriguing modal arguments about God's existence, about the nature of God, about free will, about the nature of the mind, and on and on. The history of philosophy is just full of fascinating modal arguments. But many of these arguments are difficult to understand unless one has at least a basic grasp of the logic of modality. Thus, a rudimentary understanding of modal logic opens the door to many fascinating and deep arguments from the history of philosophy--modal arguments. In addition, an understanding of modal logic deepens one's understanding of language.

Exercise 23.1
Part I.
2. ~ ◊ S
3. □ ~ S

Part II.
2. ▽ U   14. □ U v ~ □ U
3. ~ ◊ A   15. □ U v □ ~ U
6. □ ~S   17. □ G ⊃ □ ~M

8. ~◊~A   18. □ U ⊃ ~ ◊ G
9. ~◊ A   21. ~G ⊃ ~◊ G
11. ~∇ A   23. ◊ N ⊃ ∇ U

Exercise 23.2

1. a,d
b,e
c,h
f,g

2. b. T   g.F
d.F   h.F
e.F   i.T

3. b. □ A ⊃ ~◊~A   true
c. ∇ A ⊃ ~◊~A   false
e. ∇ A ⊃ ◊A v ◊ ~A   true
h. ~A ⊃ ~□ A   true
j. A ⊃ ~□ ~A   true

4. b. T   5.b T
c. T   c. T

Exercise 23.3

1,2,3,4,5,7,8,9,11,12

Exercise 23.4
Part I.
True: 1,3,5,6,7,8,9,
False: 2,4,10,11

Part II.
1. (A & B) ⊃ ◊ (A & B)
2. □ ~A & □ ~B ⊃ ◊ (A & B)
4. (~A & ~ B) ⊃ ~ ◊ (A & B)
5. □ P & ∇ Q ⊃ ◊ (P & Q)
7. □ A & □ B ⊃ ◊ (A & B)
8. □ A □ B ⊃ (A ≡ B)
9. □ ~A & □ ~ B ⊃ (A ≡ B)
10. ∇ A & ∇ B ⊃ (A ≡ B)

Supplementary Exercise 23.A

Symbolize each of the following and indicate in each case whether the sentence is true or false.

1. If P is necessarily true, then it's not possible that P is false.

2. If P is true, then it's not possible that P is false.

3. If P is noncontingent, then either it's not possible P is true or it's not possible P is false.

4. If P is false, then P is not necessarily true.

5. If P is necessarily false, then P is not possibly true.

6. If P is contingent, then P is not necessarily true.

7. If P is true, then P is contingent.

8. If P is true, then P is not necessarily false.

9. If P is true, then it P is necessarily true.

10. If P is possibly false, then P is contingent.

Answers

1. T   6. T
2. F   7. F
3. T   8. T
4. T   9. F
5. T   10. F

Chapter 24

This chapter introduces a natural deduction system for modal arguments. You will notice that the system of this chapter includes the truth-functional rules of Chapters 7-11. Happy "proofing"!

Answers

Exercise 24.1

```
(2)1.R→M
2. H&R   / MvG
3.R           Simp 2
4. M          MMP, 1, 3
5. MvG        Add 4
```

```
(3)1.~H
2.S→H
3.◊~S⊃R /R v ◊M
4.~S          MMT 1, 2
5.◊~S         Poss 4
6.R           MP 3,5
7.R v ◊M      Add 6
```

```
(7)1.□R
2.R⊃G
3.~GvM /◊M
4.R       BR 1
5.G       MP 2, 4
6.~~G     DNeg 5
7.M       DS 3,6
8.◊M      Poss 7
```

```
(10)1.G→◊M
2.H&(G&R)
3.□◊M⊃A /◊A
4.G&R     Simp 2
5.G       Simp 4
6.◊G      Poss 5
7.□◊M     P2N 1, 6
8.A       MP 3, 7
9.◊A      Poss 8
```

```
(11)1.Mv(H&W)
2.M→~G
3.□(G&S)
4.◊W⊃□B /A⊃B
```

```
5.G&S       BR 3
6.G         Simp 5
7.~~G       DNeg 6
8.~M        MMT 2,7
9. H&W      DS 1,8
10.W        Simp 9
11.◊W       Poss 10
12.□B       MP 4, 11
13.B        BR 12
14.Bv~A     Add 13
15.~AvB     Comm 14
16.A⊃B      Imp 15
```

```
(13) 1.□(A⊃B)
2.B⊃G
3.~GvS
4.~(SvR)  /◊~A
5.~S&~R     DM 4
6.~S        Simp 5
7. Sv~G     Comm 3
8.~G        DS 6,7
9.~B        MT 2, 8
10.A⊃B      BR 1
11.~A       MT 9,10
12.◊~A      Poss 11
```

```
16. 1.◊A  /◊◊A
2.◊◊A      Poss 1
```

```
17.1.◊A
2.A→◊(EvB)
3.◊(EvB)→ H  /H
4. □◊(EvB)            P2N  1, 2
5. ◊(EvB)            BR 4
6. H                 MMP 3, 5
```

```
18.□□ A  /◊□A
2. □A       BR 1
3. ◊□A      Poss 2
```

```
20.1.G→□G
2. ~G→□~G
3. ◊G / G
4. □□G      2N 1,3
5. □G       BR 4
6. G        BR 5
```

Exercise 24.2

(1) 1. □(AvB)
2. □~B
3. □A⊃□E /E
```
4.          │ □(AvB)     Reit 1
5.          │ □~B        Reit 2
6.          │ AvB        BR 4
7.          │ ~B         BR 5
8.          │ BvA        Comm 6
9.          │ A          DS 7,8
```
10. □A Nec 4-9
11. □E MP 3, 10
12. E BR 11

(3) 1. □(~R⊃S)
2. □(~SvG)
3. □[(RvG)⊃H]
4. ~□HvS / S
```
5.          │ □(~R⊃S)     Reit 1
6.          │ □(~SvG)     Reit 2
7.          │ □[(RvG)⊃H]  Reit 3
8.          │ ~R⊃S        BR 5
9.          │ ~SvG        BR 6
10.         │ (RvG)⊃H     BR 7
11.         │ S⊃G         Imp 9
12.         │ ~R⊃G        HS 8, 11
13.         │ ~~RvG       Imp 12
14.         │ RvG         DNeg 13
15.         │ H           MP 10, 14
```
16. □H   Nec 5-15
17. ~~□H   DNeg 16
18. S   DS 4, 17

(6) 1. □(H⊃S)
2. □(S⊃A)
3. □[◊(H⊃A)⊃G] /□(E⊃G)
```
4.          │ □(H⊃S)        Reit 1
5.          │ □(S⊃A)        Reit 2
6.          │ H⊃S           BR 4
7.          │ S⊃A           BR 5
8.          │ H⊃A           HS 6,7
9.          │ □[◊(H⊃A)⊃G]   Reit 3
10.         │ ◊(H⊃A)⊃G      BR 9
```

11.	◊(H⊃A)	Poss 8
12.	G	MP 10,11
13.	Gv~E	Add 12
14.	~EvG	Comm 13
15.	E⊃G	Imp 14
16. □(E⊃G)		Nec 4-15

(8)1.□(A⊃B)
2.□(B⊃~C) /□~(A&C)

3.	□(A⊃B)	Reit 1
4.	□(B⊃~C)	Reit 2
5.	A⊃B	BR 3
6.	B⊃~C	BR 4
7.	A⊃~C	HS 5,6
8.	~Av~C	Imp 7
9.	~(A&C)	DM 8
10. □~(A&C)		Nec 3-9

(11)1.□[(A&B)&R] /□A&(□B & □R)

2.	□[(A&B)&R]	Reit 1
3.	(A&B)&R	BR 2
4.	A&B	Simp 3
5.	A	Simp 4
6.	B	Simp 4
7.	R	Simp 3
8. □A		Nec 2-7
9. □B		Nec 2-7
10. □R		Nec 2-7
11. □B & □R		Conj 9,10
12. □A & (□B & □R)		Conj 8,11

(13).1.◊□A /□◊□A

| 2. | ◊□A | Reit 1 |
| 3. □◊□A | | Nec 2 |

(15)1.□A /□□A

| 2. | □A | Reit 1 |
| 3. □□A | | Nec 2 |

(18)1.A→B
2.□A /□B

3.	A→B	Reit 1
4.	□A	Reit 2
5.	A	BR 4

```
6.          |B              MMP 3,5
7. □B              Nec 3-6


(19)1.A→B
2.□~B   /□~A
3.          |A→B          Reit 1
4.          |□~B          Reit 2
5.          |~B           BR 4
6.          |~A           MMT 3, 5
7. □~A              Nec 3-6


(20)1.□A v □B
2.~□A    /□(AvB)
3.          |□A v □B      Reit 1
4.          |~□A          Reit 2
5.          |□B           DS 3,4
6.          |B            BR 5
7.          |BvA          Add 6
8.          |AvB          Comm 7
9. □(AvB)              Nec 3-8
```

Exercise 24.3

```
(2)1.A ↔ G
2. ~G
3. ~A⊃□E / E
4.□(A≡G)              Double Arrow Ex 1
5.A≡G              BR 4
6.(A⊃G)&(G⊃A)     Equiv 5
7.A⊃G              Simp 6
8.~A              MT 7, 2
9.□E              MP 3, 8
10.E              BR 9


(5) 1.A ⊃□B
2. ~B /◊~A
 3. ◊~B           Poss 2
4. ~□B            DE 3
5. ~A             MT 1,4
6. ◊~A            Poss 5


(7)1.A ↔ B
 2.B→G / A→G
3. (A→B)&(B→A)              ME 1
4. (A→B)              Simp 3
```

```
5.  (A→G)                    MHS 2, 4


(9) 1. A↔B
2.  □~A    /  □~B
3.  (A→B)&(B→A)              ME 1
4.  (B→A)                    Simp 3
5.          (B→A)            Reit 4
6.          □~A              Reit 2
7.          ~A               BR 6
8.          ~B               MMT 5,7
9. □~B                       Nec 5-8


(10)1. A↔B
2. ~A  /~B
3. (A→B) & (B→A)             ME 1
4. (B→A)                     Simp 3
5. ~B                        MMT 2,4


(12)1. A→B
2. B⊃D   /A⊃D
3. □(A⊃B)                    Arrow Ex 1
4. A⊃B                       BR 3
5. A⊃D                       HS 2,4

(13)1. ~◊A  / ◊~A
2. □~A                       DE 1
3. ~A                        BR 2
4. ◊~A                       Poss 3
```

Exercise 24.4

```
(2)1. □(A⊃B)
2. ◊A / ◊B
3.        | ~◊B         AP
4.        | □~B         DE 3
5.        |      | □~B        Reit 4
6.        |      | □(A⊃B)     Reit 1
7.        |      | ~B         BR 5
8.        |      | A⊃B        BR 6
9.        |      | ~A         MT 7,8
10.       | □~A         Nec 5-9
11.       | ~ ◊A       DE 10
12.       |    ◊A&~◊A     Conj 2,11
13. ◊B         IP 3-12
```

(4) 1. ◊(AvB)   /◊A v ◊B
2. | ~(◊A v ◊B)   AP
3. | ~◊A & ~◊B   DM 2
4. | □~A & □~B   DE 3
5. | □~A   Simp 4
6. | □~B   Simp 4
7. | | □~A   Reit 5
8. | | □~B   Reit 6
9. | | ~A   BR 7
10. | | ~B   BR 8
11. | | ~A & ~B   Conj 9, 10
12. | | ~(A v B)   DM 11
13. | □~(A v B)   Nec 7-12
14. | ~◊(A v B)   DE 13
15. | ◊(A v B) & ~◊(A v B) Conj 1,14
16. ◊A v ◊B   IP 2-15

(5) 1. □(A ⊃ B)
2. ◊~B   /◊~A
3. | ~◊~A   AP
4. | □A   DE 3
5. | | □A   Reit 4
6. | | □(A⊃B)   Reit 1
7. | | A   BR 5
8. | | A⊃B   BR 6
9. | | B   MP 7,8
10. | □B   Nec 5-9
11. | ~□B   DE 2
12. | □B & ~□B   Conj 10,11
13. ◊ ~A   IP 3-12

Exercise 24.5

2.   1. A→(D&T)
      2. ~◊D/~◊A
      3. | A→(D&T)   Reit 1
      4. | ~◊D   Reit 2
      5. | □[A⊃(D&T)]   Arrow Ex, 3
      6. | A⊃(D&T)   BR 5
      7. | ~Av(D&T)   Imp 6
      8. | (~AvD)&(~AvT)   Dist 7
      9. | ~AvD   Simp 8
      10. | □~D   DE 4
      11. | ~D   BR, 10
      12. | Dv~A   Comm 9

107

```
     13.        | ~A              DS 11,12
     14.□~A                       Nec 3-13
     15.~◊A                       DE 14



3.   1.SvP
     2.R⊃~◊P
     3.~S/~R
       ‾‾‾‾‾
     4.P              DS 1,3
     5.◊P             Poss 4
     6.~~◊P           DNeg 5
     7.~R             MT 2,6


4.   1.~◊G
     2.L⊃G
     3.~L⊃[(M&A)&(R&P)]/P
       ‾‾‾‾‾‾‾‾‾‾‾‾‾‾‾‾‾‾‾
     4.□~G              DE 1
     5.~G               BR 4
     6.~L               MT 2, 5
     7.(M&A)&(R&P)      MP 3, 6
     8.R&P              Simp 7
     9.P                Simp 8
```

G:	Glenn and Larry occupy ...
L:	Glenn and Larry both deliver ...
M:	Melissa and Val ...
A:	Alan and Charlie ...
R:	Ron and Mel ...
P:	Patty will be disappointed.

```
(7)1.~◊(A & D)
   2. ◊D→◊E    /□(~A v ~D)
   3.        | ~◊(A & D)        Reit 1
   4.        | □~(A & D)        DE 3
   5.        | ~(A & D)         BR 4
   6.        | ~A v ~D          DM 5
   7. □(~A v ~D)                Nec 3-6


(8)1.□(A→D)
   2. □(D→B)    /□(~B→~A)
   3.        | □(A→D)           Reit 1
   4.        | □(D→B)           Reit 2
   5.        |  (A→D)           BR 3
   6.        |  (D→B)           BR 4
   7.        |  (A→B)           MHS 5,6
```

```
8.          │□ (A⊃B)     Arrow Ex 7
9.          │□ (~B⊃~A)   Trans 8
10.         │~B→~A       Arrow Ex 9
11. □(~B→~A)             Nec 3-10


(9) 1.◊A→◊G
2. A
3. ◊H→~◊G   /~◊ H
4.          │~~◊H        AP
5.          │◊H          DNeg 4
6.          │~◊G         MMP 3,5
7.          │~◊A         MMT 1,6
8.          │◊A          Poss 2
9.          │◊A & ~◊A    Conj 7,8
10. ~◊H                  IP 4-9


(13)1. T→□T
2. ~T→□~T
3. T
4. □T→~W
5. ~W→~R    /~R
6. □T                    MMP 1,3
7. ~W                    MMP 4,6
8.~R                     MMP 5,7


(14)1.~◊(G & E)
2. E   /~G
3. □~(G & E)             DE 1
4. ~(G & E)              BR 3
5. ~G v ~E              DM 4
6. ~ ~E                 DNeg 2
7. ~E v ~G              Comm 5
8. ~G                   DS 6,7
```

Exercise 24.6

```
2.   1.          │AvB           AP
     2.          │BvA           Comm 1
     3.          │~~BvA         DNeg 2
     4.          │~B⊃A          Imp 3
     5.     (AvB)⊃(~B⊃A)        CP 1-4
     6.     □[(AvB)⊃(~B⊃A)]     Taut Nec 5

3.   1.          │B             AP
     2.          │Bv~A          Add 1
```

```
        3           |~AvB              Comm 2
    4.              |A⊃B               Imp 3
    5.        B⊃(A⊃B)                  CP 1-4
    6.        □[B⊃(A⊃B)]               Taut Nec 5


6.  1.              |(A⊃B)&~B          AP
    2.              |A⊃B               Simp 1
    3.              |~B                Simp 1
    4.              |~A                MT 2, 3
    5.        [(A⊃B)&~B]⊃~A            CP 1-4
    6.        □{[(A⊃B)&~B]⊃~A}         Taut Nec 5


8.  1.              |[(A⊃B)&(B⊃R)]          AP
    2.              |A⊃B                    Simp 1
    3.              |B⊃R                    Simp 1
    4.              |A⊃R                    HS 2, 3
    5.        [(A⊃B)&(B⊃R)]⊃(A⊃R)           CP 1-4
    6.        □{[(A⊃B)&(B⊃R)]⊃(A⊃R)}        Taut Nec 5


9.  1.              |[(A⊃B)&(R⊃S)]&(AvR)        AP
    2.              |(A⊃B)&(R⊃S)               Simp 1
    3.              |A⊃B                       Simp 2
    4.              |R⊃S                       Simp 2
    5.              |AvR                       Simp 1
    6.              |BvS                       CD 3, 4, 5
    7.        {[(A⊃B)&(R⊃S)]&(AvR)}⊃(BvS)      CP 1-6
    8.        □{{[(A⊃B)&(R⊃S)]&(AvR)}⊃(BvS)}   Taut Nec 7

11. 1.              |[(A&B)&(A⊃R)]     AP
    2.              |A&B               Simp 1
    3.              |A⊃R               Simp 1
    4.              |A                 Simp 2
    5.              |R                 MP 3, 4
    6.        [(A&B)&(A⊃R)]⊃R          CP 1-5
    7.        □{[(A&B)&(A⊃R)]⊃R}       Taut Nec 6

12. 1.              |A         AP
    2.              |Av~A      Add 1
    3.              |~AvA      Comm 2
    4.              |A⊃A       Imp 3
    5.        A⊃(A⊃A)   CP 1-4
    6.        □[A⊃(A⊃A)]      Taut Nec 5
```

110

```
14.   1.          │ AvB            AP
      2.          │ BvA            Comm 1
      3.          │ ~~BvA          DNeg 2
      4.          │ ~B⊃A           Imp 3
      5.     (AvB)⊃(~B⊃A) CP 1-4
      6.     □{(AvB)⊃(~B⊃A)} Taut Nec 5

18.   1.          │ □A                 AP
      2.          │    │ □A            Reit 1
      3.          │    │ A             BR, 2
      4.          │    │ AvB           Add 3
      5.          │ □(AvB)             Nec 2-4
      6. □A⊃□(AvB)                     CP 1-5
      7.  □[□A⊃□(AvB)]                 Taut Nec 6
      8. □A→□(AvB)                     Arrow Ex 7
19.   1.          │ A             AP
      2.          │ AvB           Add 1
      3.     A⊃(AvB)           CP 1-2
      4.     □[A⊃(AvB)]        Taut Nec 3
      5.     A→(AvB)           Arrow Ex 4

21.   1.          │ A&B           AP
      2.          │ A             Simp 1
      3.          │ AvB           Add 2
      4.     (A&B)⊃(AvB)         CP 1-3
      5.     □[(A&B)⊃(AvB)]      Taut Nec 4

22.   1.          │ ◊~A           AP
      2.          │ ~□A           DE 1
      3.     ◊~A ⊃ ~□A           CP 1-2
      4.     □[◊~A ⊃ ~□A]        Taut Nec 3
      5.     ◊~A→~□A             Arrow Ex 4

23.   1.          │ □~A           AP
      2.          │ ~◊A           DE 1
      3.     □~A⊃~◊A             CP 1-2
      4.     □[□~A⊃~◊A]          Taut Nec 3
      5.     □~A→~◊A             Arrow Ex 4

24.   1.          │ ~(◊Av◊~A)         AP
      2.          │ ~◊A&~◊~A          DM 1
      3.          │ □~A&□A            DE 2
      4.          │ □~A               Simp 3
      5.          │ □A                Simp 3
```

6.		~A	BR 4
7.		A	BR 5
8.		A&~A	Conj 6,7
9.	◊Av◊~A		IP 1-8

27.
1.		(A→B)	AP
2.		□(A⊃B)	Arrow Ex 1
3.	(A→B)⊃□(A⊃B)		CP 1-2
4.	□[(A→B)⊃□(A⊃B)]		Taut Nec 3
5.	(A→B)→□(A⊃B)		Arrow Ex 4

29.
1.		A→B	AP
2.		□(A⊃B)	Arrow Ex 1
3.		□(~AvB)	Imp 2
4.		□~(~~A&~B)	DM 3
5.		□~(A&~B)	DNeg 4
6.		~◊(A&~B)	DE 5
7.	(A→B)⊃~◊(A&~B)		CP 1-6
8.	□[(A→B)⊃~◊(A&~B)]		Taut Nec 7
9.	(A→B)→~◊(A&~B)		Arrow Ex 8

31.
1.	A&~B		AP
2.		~~(A→B)	AP
3.		A→B	DNeg 2
4.		A	Simp 1
5.		B	MMP, 3,4
6.		~B	Simp 1
7.		B&~B	Conj 5,6
8.	~(A→B)		IP 2-7
9.	(A&~B)⊃~(A→B)		CP 1-8
10.	□[(A&~B)⊃~(A→B)]		Taut Nec 9
11.	(A&~B)→~(A→B)		Arrow Ex 10

32.
1.		~~◊(A&~A)	AP
2.		◊(A&~A)	DNeg 1
3.		~□~(A&~A)	DE 2
4.			A&~A	AP
5.			A	Simp 4
6.			~A	Simp 4
7.			A&~A	Conj 5,6
8.		~(A&~A)	IP 4-7
9.		□~(A&~A)	Taut Nec 8
10.		~◊(A&~A)	DE 9
11.		◊(A&~A) & ~◊(A&~A)	Conj 2,10
12.	~◊(A&~A)		IP 1-11

33.   1. | (A↔B)   AP
      2. | □(A≡B)   Double Arrow Ex, 1
      3.   (A↔B)⊃□(A≡B)   CP 1-2
      4.   □[(A↔B)⊃□(A≡B)]   Taut Nec 3
      5.   (A↔B)→□(A≡B)   Arrow Ex 4

34.   1. | ◊(A↔B)   AP
      2. | A↔B   AP
      3. | (A↔B)⊃(A↔B)   CP 2
      4. | □[(A↔B)⊃(A↔B)]   Taut Nec 3
      5. | (A↔B)→(A↔B)   Arrow Ex 4
      6. | ◊(A↔B)   P2N 1, 5
      7. | (A↔B)   BR 6
      8. | □(A≡B)   Double Arrow Ex 7
      9. | (A≡B)   BR 8
   10.   ◊(A↔B)⊃(A≡B)   CP 1-5
   11.   □[◊(A↔B)⊃(A≡B)]   Taut Nec 10
   12.   ◊(A↔B)→(A≡B)   Arrow Ex 11

37.   1. | ◊(A→B)   AP
      2. | A→B   AP
      3. | (A→B)⊃(A→B)   CP 2
      4. | □[(A→B)⊃(A→B)]   Taut Nec 3
      5. | (A→B)→(A→B)   Arrow Ex 4
      6. | □(A→B)   P2N, 1, 5
      7. | A→B   BR 6
      8.   ◊(A→B)⊃(A→B)   CP 1-7
      9. □[◊(A→B)⊃(A→B)]   Taut Nec 8
   10. ◊(A→B)→(A→B)   Arrow Ex 9

Exercise 24.7

(2)1.□□◊A   /◊A
 2. ◊A   Red 1

(3)1.◊◊A   /◊A
 2.◊A   Red 1

(6)1.◊□◊P v □Q
2. ◊~Q
3. ◊P⊃□P / P

113

```
4.  ◊P v □Q              Red 1
5.  ~□Q                  DE 2
6.  □Q v ◊P              Comm 4
7.  ◊P                   DS 5,6
8.  □P                   MP 3,7
9.  P                    BR 8


(7) 1. □◊◊□P
2.  Q⊃~P
3. ~□Q⊃□S / S
4.  □P                   Red 1
5.  P                    BR 4
6.  ~~P                  DNeg 5
7.  ~Q                   MT 6,2
8.  ◊~Q                  Poss 7
9.  ~□Q                  DE 8
10. □S                   MP 3,9
11. S                    BR 10


(9) 1. ◊◊◊◊□P
2.  P→Q
3.  Q→R   /R
4.  □P                   Red 1
5.  P                    BR 4
6.  P→R                  MHS 2,3
7.  R                    MMP 5,6


(10) 1. □◊◊P
2. □S→□~P
3. ◊~S→□G /G
4.  ◊P                   Red 1
5.  ~□~P                 DE 4
6. ~□S                   MMT 2,5
7.  ◊~S                  DE 6
8.  □G                   MMP 3,7
9.  G                    BR 8
```

Chapter 25

Deductive arguments are not the only arguments in town. Many arguments in everyday life are *inductive* in nature. For instance, arguments in the courtroom are typically inductive, arguments in politics are usually inductive, and the arguments people make in the ordinary business of life are typically inductive. This chapter covers three very important types of inductive reasoning:

1. analogical reasoning
2. enumerative induction
3. inference to the best explanation

Whether you realize it or not, you frequently reason in accord with the general patterns discussed in this chapter.

Analogical arguments. Many arguments in everyday life are analogical. This chapter teaches the logical structure of analogical reasoning and introduces techniques of evaluation. It can be useful to think up examples from everyday life and then discuss these with others in class. We reason better when we understand the nature of our reasoning, and just becoming aware of the logical structure of analogical reasoning can help us improve our reasoning skills.

Enumerative Induction and Inference to the Best Explanation Similarly, many arguments in everyday life are enumerative inductions and many others are inferences to the best explanation. Again, it can be useful to think up examples from everyday life and then discuss these with others in class. This chapter teaches the logical structure of these important forms of inductive reasoning, along with techniques of evaluation. Again, we reason better when we understand the nature of our reasoning, and just learning the logical structure of these inductive arguments can improve our reasoning abilities.

Answers

Exercise 25.1

1.a. One is a Republican, the other is a Democrat. One is Catholic, the other is Baptist.
b. Both have family histories involving heart attacks. Neither exercises.
c. Pete will have a heart attack very soon.
d. Pete will have a heart attack or a heart condition.
e. Pete has been having heart pains lately.
f. Although Joe's family has a history of heart attacks, Pete's family has no past heart attacks in its history.

3. a. weaker	5. b. decreases
b. weaker	d. unaffected
c. stronger	f. increases
d. stronger	
e. stronger	6. a. decreases
f. weaker	b. increases
g. unaffected	c. decreases

h. stronger

4.a. weaker
b. weaker
c. unaffected
d. unaffected
e. stronger
f. stronger
g. stronger

d. increases
e. unaffected

8. a. The new buyer graduated with a 4.0.
b. The new buyer forged his credentials--he did not actually graduate from the UW fish-buying program.

Exercise 25.2

1. b. strengthens
d. weakens
f. weakens

2.a. weakens
b.strengthens
c.weakens
d. weakens
e. strengthens
f. weakens
g. weakens
h. weakens
i. strengthens

4.a. weakens
b. strengthens
c.unaffected
d. weakens
e. strengthens
f. strengthens
g. weakens
h. weakens

5. b. weakens
d. unaffected
f. unaffected
g. strengthens
i. weakens
j. weakens

7.a. weakens
b. strengthens
c. weakens
d. weakens
e. weakens
f. weakens
g. weakens
h. weakens
i. unaffected

8. b. weakens
d. strengthens
e. strengthens
g. weakens
h. unaffected
j. strengthens

k. strengthens
l. unaffected
m. weakens
n. strengthens
o. strengthens
p. strengthens

ANY ADVICE BEFORE I HEAD OFF TO LOGIC CLASS MOM?

JUST MIND YOUR P's AND Q's DEAR ...AND KEEP WARM.

Supplementary Exercise

Just in case you would like some additional problems to practice, here are some "supplementary" exercises. You might also use these as tests to make sure you are really understanding the concepts. Answers are provided below.

1. Consider this argument:

Heavy-metal music is philosophical music. I listened to three songs by Black Sabbath and all three songs had philosophical lyrics.

In each of the following cases, would the addition make the argument stronger or weaker? Or would it leave things unaffected?

a. In addition, I've also listened to songs by four other heavy metal groups and all contained philosophical lyrics.
b. Furthermore, I've also found philosophical lyrics in several soft-rock songs.
c. Also, I found seven heavy metal songs by KISS that do not contain philosophical lyrics.
d. In addition, I found seven heavy metal songs by three different groups that do not contain philosophical lyrics.
e. Additionally, the last time I played a Led Zeppelin song on my stereo, I blew out both speakers.

2. You asked a friend for a favor three times and all three times he helped you. You are about to ask for another favor. You expect that he will help you this time too. In each of the following cases, would the addition make your argument stronger or weaker? Or would it leave things unaffected?

a. The first three favors were for little things; this favor is for a big thing.
b. Before, your friend was single and had lots of time on his hands. Now, he is married and is very busy.
c. You've gotten a new job since the last time you asked for a favor.
d. The first three favors were for big things. This is a small favor.
e. Your friend has just joined a new church, The Church of Those who Grant Favors.
f. The first favor was a big favor, the second was a little one, and the third favor required that your friend make a great sacrifice for you.

3. Consider this argument:

I've liked every Gremlin album that I've bought so far. I'll probably like their next album.

In each of the following cases, would the addition make the argument stronger or weaker? Or would it leave things unaffected?

a. I've bought only two albums but they've made six.
b. I've bought five of their six albums.

c. The next album is supposed to be a whole new type of music--very much unlike any of their previous albums.

d. I've gotten a new stereo since I bought their last album.

e. Their next album won't be released until three years from now.

4. Consider this argument.

I've eaten three times at Tiny's Burger stand, and I liked the food each time. I'll probably like the food the next time I eat there.

In each of the following cases, would the addition make the argument stronger or weaker? Or would it leave things unaffected?

a. The first time I ate a burrito, the second time I ate a hamburger, the third time I ate a fishburger.

b. All three previous times I ate hamburgers, this time I'm going to eat a fishburger.

c. On all three previous visits, the meals were cooked by Tiny himself. The next trip, Tiny will be on vacation and the food will be cooked by his manager, Wimpy.

d. I've actually only eaten there twice.

e. The next time I eat at Tiny's, I'm going to try something new: his "oyster-burger."

f. All three previous times I ate hamburgers, this time I'm going to eat a hamburger again.

g. Since my last visit, I've become a vegan.

5. Consider this argument:

I've met four professional athletes and all four were polite and well-mannered individuals. Therefore, the next professional athlete I meet will probably also be polite and well-mannered.

In each of the following cases, would the addition make the argument stronger or weaker? Or would it leave things unaffected?

a. All four were baseball players.

b. One was a baseball player, one played basketball, one played football, and one played soccer.

c. One of the four was from L.A., one was from New York, one was from Seattle, and one was from Alabama.

d. All four were from Seattle.

e. All four were in their thirties.

f. One was still in his teens, one was in his twenties, one was in his thirties, and one was over 40.

g. I've actually met 8, and all were well-mannered.

Answers

1. a. Stronger
   b. Unaffected
   c. Weaker
   d. Weaker
   e. Unaffected

2. a. Weaker
   b. Weaker
   c. Unaffected
   d. Stronger
   e. Stronger
   f. Stronger

3. a. weaker
   b. Stronger
   c. Weaker
   d. Unaffected
   e. Weaker

4. a. Stronger
   b. Weaker
   c. Weaker
   d. Weaker
   e. Weaker
   f. Stronger
   g. Weaker

5. a. Weaker
   b. Stronger
   c. Stronger
   d. Weaker
   e. Weaker
   f. Stronger
   g. Stronger

Chapter 26

This chapter treats two main issues. First, what patterns of reasoning are employed by scientists when they "prove" or "disprove" scientific theories? What is the "logic" of science? Second, what patterns or forms of reasoning are used when scientists track down the *causes* of things?

More specifically, this chapter:

1. explains the basic logic of the scientific method.
2. explains, in a rudimentary way, the concept of "cause."
3. introduces Mill's methods.

Science affects us all. Whether we like it or not, science has become an important part of modern life. After studying this chapter, you should have a deeper understanding of the *logic* of science.

Answers

Exercise 26.2

2. Concommitant variation	13. Difference and Agreement
3.Difference	14. Agreement
4. Difference	17.Difference
8. Difference	18.Agreement and difference
9. Difference	19.Agreement

Exercise 26.3

1. Ask each athlete, separately, "Why did you choose Smith's class? See if there is a common reason.

2. Examine all the situations in which stuttering attacks occur. List the characteristics of each situation. See if there is a common condition or factor occurring in each case of stuttering.

5. Make a list of what each person ate and see if there are any foods that were common to all cases of illness.

6. Identify the factors common to the three surviving bushes. Check to see if these are absent in the cases of the three bushes that died.

10. Keep a record of which barista makes each day's latte. See if the good lattes are all made by the same barista; or, perhaps the lousy lattes are all made by the same barista.

11. Make a list of the factors which are common to all of your busy periods.

Appendix 1

This appendix contains truth-functional tree-tests for:

1. Logical status of a sentence.

2. Validity

The problems of Appendix 1 are answered below, *except for* the following problems, which were left unanswered in case your teacher wishes to assign them as homework:

Exercise A.1: problems 12-15.
Exercise A.2: problems 11-14.
Exercise A.3: problems 10-14.

Exercise A.1

In the diagrams that follow, I left the check marks off so that the trees are not as cluttered.

T: Tautology. C: Contradiction. X: Contingent

(6)
```
    │ P ∨ (P & ~P)
    │ P
    │ P & ~P        (X)
  ┌─┴──┐
  P    ~P
    P
    x
```

```
 P ∨ (P & ~P)│
P │      P & ~P
           P
           ~P
              P
              x
```

(7)
```
    │ P ⊃ (P ∨ Q)
  P │ P ∨ Q
    │ P
    │ Q        (T)
    x
```

(8)
```
        │ (P & ~P) ⊃ Q
 P & ~P  │ Q
    P   │
   ~P   │        (T)
        │  P
    x
```

(9) ~(P ∨ ~P)│
```
           │ P ∨ ~P
           │ P
           │ ~P
  (C)      │
        P  │
           x
```

(10) ~[P ⊃ (Q ⊃ P)]│
```
            │ P ⊃ (Q ⊃ P)
       P    │ Q ⊃ P
       Q    │ P
            │        (C)
       x
```

(11) P ≡ ~P│
```
       ┌────┴────┐
    P          P
   ~P         ~P
     P    (C)   P
     x          x
```

122

Exercise A.2
In the diagrams that follow, I left the check marks off so that the trees are not as cluttered.

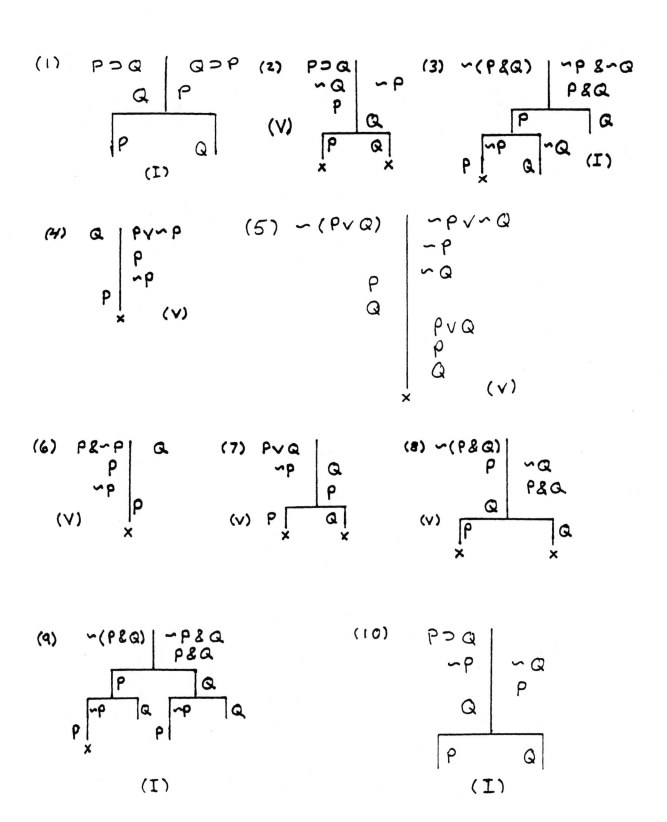

123

Exercise A.3
In the diagrams that follow, I left the check marks off so that the trees are not as cluttered.

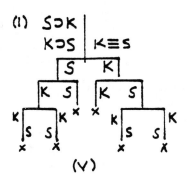

(1) S⊃K │
   K⊃S │ K≡S
      S │ K
  K S │ K S
 K │ K × × │ K
 S │ S   S │ S
 × │ ×   × │ ×
    (∨)

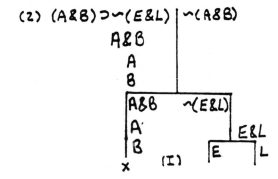

(2) (A&B)⊃∼(E&L) │ ∼(A&B)
  A&B
   A
   B
  A&B   ∼(E&L)
  A │   E&L
  B │ E │ L
  × (I)

(3) E∨(A&C) │ (E∨A)&(E∨C)
   E │   A&C
 E∨A │ E∨C   A
 E │ E   C
 A │ C   E∨A │ E∨C
 ×   ×   E │ E
    (∨)   A │ C
      ×   ×

(4) G⊃(C⊃H) │ (G&C)⊃H
  G&C │ H
   G
   C
  G │ C⊃H
  × │ C │ H
  (∨)   ×   ×

(5) E │ E∨A
     E
     A
    ×
  (∨)

(6) J │
  M │ J&M
 J │ M
 ×   ×
  (∨)

(7) F⊃S │ ∼F∨S
     ∼F
     S
  F
 F │ S
 ×   ×
  (∨)

(8) I⊃J │ ∼J⊃∼I
  ∼J │ ∼I
   I │
     J
 I │ J
 ×   ×
  (∨)

(9) W⊃M │
  D⊃Y │
  W∨D │ M∨Y
      M
      Y
    W │ M
  D │ Y ×
     ×
 W │ D │
 ×   × (∨)

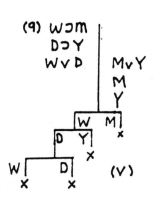

124